Lecture Notes in Applied and Computational Mechanics

Volume 74

About this Series

This series aims to report new developments in applied and computational mechanics—quickly, informally and at a high level. This includes the fields of fluid, solid and structural mechanics, dynamics and control, and related disciplines. The applied methods can be of analytical, numerical and computational nature.

More information about this series at http://www.springer.com/series/4623

Thomas Lenarz · Peter Wriggers
Editors

Biomedical Technology

Springer

Editors
Thomas Lenarz
Department of Otorhinolaryngology
Hannover Medical School
Hannover
Germany

Peter Wriggers
Institute of Continuum Mechanics
Leibniz Universität Hannover
Hannover
Germany

ISSN 1613-7736 ISSN 1860-0816 (electronic)
ISBN 978-3-319-10980-0 ISBN 978-3-319-10981-7 (eBook)
DOI 10.1007/978-3-319-10981-7

Library of Congress Control Number: 2014952593

Springer Cham Heidelberg New York Dordrecht London

Printed on acid-free paper

Springer is part of Springer Science+Business Media (www.springer.com)

Preface

One of the current challenges in medicine and engineering is related to the application of computational methods to clinical medicine. The virtual environment can be used to study biological systems at different scales and under multi-physics conditions. Based on the tremendous advances in medical imaging, modern CAD systems, and high-performance computing, engineering can provide help in understanding biological processes but also implant designs. This enables the possibility to enhance medical decision processes in many areas of clinical medicine. The computational tools and methods can be applied to predict performance of medical devices in virtual patients. Physical and animal testing procedures can be reduced by use of virtual prototyping of medical devices.

In this book, scientists from scientists from different areas of medicine, engineering, and natural sciences have contributed to the above research areas and ideas. The book will focus on function, production, initialization, and complications of different types or implants and related topics.

The contributions start with theoretical and numerical investigations that are related to modeling biological materials like the papers "RVE Procedure for Estimating the Elastic Properties of Inhomogeneous Microstructures Such as Bone Tissue" by Blöß and Welsch and "A Gradient-Enhanced Continuum Damage Model for Residually Stressed Fibre-Reinforced Materials at Finite Strains" by Waffenschmidt et al. A more application-oriented work "A Mechanically Stimulated Fracture Healing Model Using a Finite Element Framework" is provided by Sapotnick and Nackenhorst that buidls a bridge to the work "The Customized Artificial Hip Cup: Design and Manufacturing of an Innovative Prosthesis" by Betancur Escobar et al. New stents are modeled in the paper "On the Role of Phase Change in Modelling Drug-Eluting Stents" by Bozsak et al. The paper "Development of Magnesium Alloy Scaffolds to Support Biological Myocardial Grafts: A Finite Element Investigation" by Weidling et al. deals with the development of new degenerative implants. The contributions "Finite Element Analysis of Transcatheter Aortic Valve Implantation in the Presence of Aortic Leaflet Calcifications" by Dimasi et al., "Repair of Mitral Valve Prolapse Through ePTFE Neochordae: A Finite Element Approach From CMR" by Sturla et al. and

"An Extended Computational Framework to Study Arterial Vasomotion and Its Links to Vascular Disease" by Boileau et al. are related to virtual models for the vascular system. Models that describe the behavior of the cochlea are provided in "Development of a Model of the Electrically Stimulated Cochlea" by Nogueira et al. Finally, models and investigations of infections due to implantation are discussed in "Implant Related Infections" by Abraham and "Animal Test Models for Implant-Associated Inflammation and Infections" by Rais et al.

All contributions show the state of the art in modeling and numerical simulation of systems in biotechnology and thus provide an extensive overview of this subject.

Hannover, May 2014

Thomas Lenarz
Peter Wriggers

Contents

RVE Procedure for Estimating the Elastic Properties of Inhomogeneous Microstructures Such as Bone Tissue

Tanja Blöß and Michael Welsch

Abstract Cancellous bone can roughly be seen as a two-phase material consisting of the bone tissue reinforcement and the interstitial bone marrow matrix. Thus, for a computer-aided mechanical stress analysis of bones a constitutive law is required, which can predict the inhomogeneous elasticity depending on the local bone density and microstructure. Besides several measurement methods, the method of representative volume element (RVE) in combination with the finite element solution technique has been established for this purpose. This work investigates this method in detail. Therefore, random but statistical equivalent RVEs are created to have unlimited access to different structures. Generally, an apparent and not an effective stiffness is obtained due to the RVE method. However, a very close solution can be achieved if several issues are considered carefully. These issues can be divided into the set of boundary conditions, the RVE size and averaging the randomness. The influences are investigated accurately. A new approach is proposed to deduce an isotropic constitutive law from the anisotropic stiffness matrix. There are unlimited possible solutions in theory. However, the Voigt and Reuss approximations give the possible bounds. A method is described, which allows to obtain the effective stiffness by merging these bounds. A structural analysis is performed with different RVEs and the effective stiffness is estimated for varying parameters. An empirical equation is introduced, which covers the whole stiffness range. Therein, the microstructure is modelled with a single parameter. Real bone measurements can be fitted with this equation as well.

Keywords Cancellous bone · Boundary conditions · Elastic properties · FEM · Homogenization · Voigt and Reuss approximation

T. Blöß (✉) · M. Welsch
Institute for Mechanical Engineering and Computer-Assisted Product Development, Helmut Schmidt University/Bundeswehr University, Hamburg, Germany
e-mail: tanja.bloess@hsu-hh.de

M. Welsch
e-mail: michael.welsch@hsu-hh.de

T. Lenarz and P. Wriggers (eds.), *Biomedical Technology*, Lecture Notes in Applied and Computational Mechanics 74, DOI 10.1007/978-3-319-10981-7_1

1 Introduction

Computer simulations become more and more important for endoprosthetic investigations of bones. Therefore, a realistic material modeling is required to ensure a reliable prediction of the inner mechanical stresses. Bones generally consist of cancellous bone surrounded by a thin layer of dense compact bone resulting in location-dependent material properties. The modelling of the microstructure in detail is computational out of reach nowadays. A pointwise homogenization of the stochastic and heterogeneous microstructure would be beneficial. Thus, a constitutive law is required that can predict the inhomogeneous elasticity depending on the local bone density and microstructure.

Direct mechanical measurements for example are performed by Ashmann et al. [1, 2], Rho et al. [3, 4], Dalstra et al. [5] and different regression equations are proposed. In the mid nineties a new idea was investigated. Real microstructures based on high resolution CT images are converted into virtual models that could be studied by FEM simulation. Such studies are performed by Müller [6], Ulrich [7], Pahr and Zysset [8]. Different issues raise by dealing with the continuum mechanics approach. Ulrich et al. investigated the influence of meshing and element formulation. Pahr and Zysset compared several sets of boundary conditions regarding the accuracy of the obtained stiffness of human cancellous bone specimens.

Since the procedure of calculating the anisotropic stiffness matrix seems to be clear, it lacks of estimating a corresponding isotropic constitutive law. The theory of micromechanics and homogenization points out to distinguish between apparent and effective estimates. As a general rule, an apparent estimate is obtained since the window size is limited. However, a convergence study allows the prediction of an effective estimate by increasing the window size stepwise (cp. Kanit [9, 10]).

Notwithstanding that the FEM solution is an approximation by nature, an apparent estimate should be expected generally due to use of boundary conditions.

This work presents a study of the different influences and proposes a procedure to calculate effective moduli. Methods are presented to determine the "effectiveness" of the solution. Plenty different, but stochastically equivalent structures are needed to study the influences entirely. An algorithm is applied to generate an unlimited number of varying representative volume elements (RVE).

2 Material and Method

2.1 Generation of Stochastic RVE

Cancellous bone can roughly be seen as a two-phase material consisting of the bone tissue reinforcement and the interstitial bone marrow matrix. Generally, the effective elastic properties of such materials are depending on the respective volume fractions, the elastic properties of both materials and the structural composition.

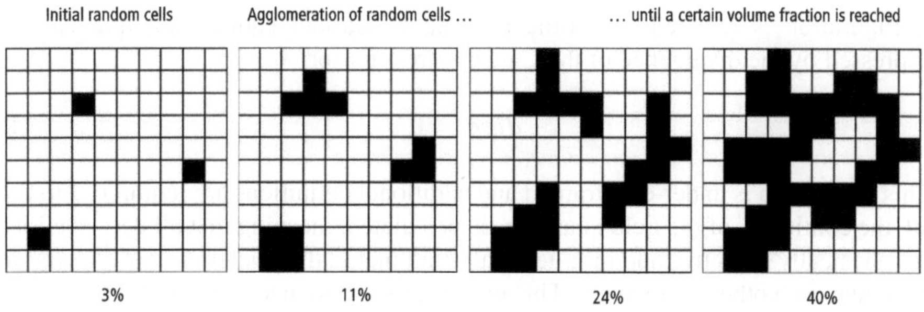

Fig. 1 Illustration of the RVE generating algorithm. Material is agglomerated randomly around existing material

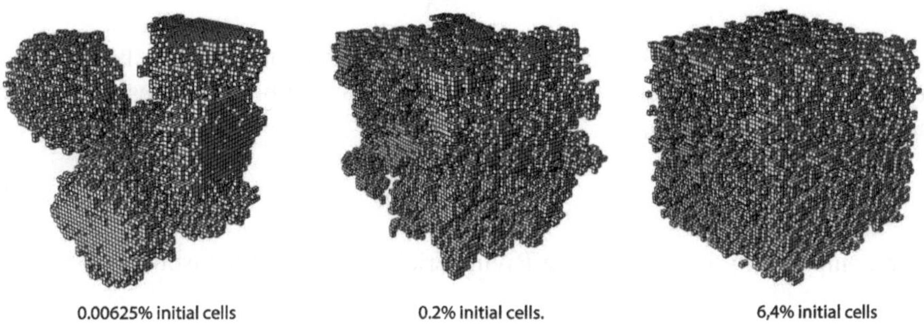

Fig. 2 Agglomerated material of different structures is emerged from varying number of initial cells. *Left* 0.00625 %, *middle* 0.2 % and *right* 6.4 % initial cells, all with 25 % volume fraction in a grid of $50 \times 50 \times 50$

A simple algorithm generates random RVE structures in three steps. First an initial number of cells are randomly assigned with material within a three-dimensional grid. Afterwards additional cells are randomly selected, but only assigned with material, if they are adjacent to existing material. This is done until a given volume fraction is reached. In a final step, all remaining grid cells are assigned with the matrix material. This algorithm is illustrated in Fig. 1.

While the volume fraction is directly regulated, the RVE structure develops indirectly by the number of initial cells. A rather rough cluster with high material agglomerations emerges from a small number of initial cells, whereas a fine dispersed cluster emerges from many initial cells. Figure 2 shows three different RVEs, each with 50 elements per edge and equal volume fraction but varying number of initial cells.

2.2 Continuum Mechanics Approach

The continuum mechanics approach is used to calculate three-dimensional material deformations. For static considerations the momentum balance of the current configuration is reduced to a time and mass invariant equilibrium that can be expressed by the divergence of the Cauchy stress tensor.

$$div\boldsymbol{\sigma} = 0 \tag{1}$$

This expression is under-constrained and additional definitions are required. First of all, the continuity of the field quantities is postulated, meaning that all deformations are physically objective and infinitesimal small material particles are not allowed to penetrate each other or fluctuate. This uniqueness is obtained by the definition of the deformation tensor.

$$\mathbf{F} = \frac{dx}{dX} \tag{2}$$

In terms of the physical objectivity, the material behavior of elastic bodies (also denoted as Cauchy elasticity) now demand tensor compatibility of stress and deformation.

$$\boldsymbol{\sigma} = f(\mathbf{F}) \tag{3}$$

Such compatibility is given by the Rivlin-Ericksen theorem for isotropic behavior [11].

$$\boldsymbol{\sigma} = a_0 + a_1\mathbf{F}^T\mathbf{F} + a_2(\mathbf{F}^T\mathbf{F})^2 \tag{4}$$

Thereby, the coefficients a_0, a_1 and a_2 are arbitrary scalar functions of the invariants of the deformation tensor. Usually, this dependence is formulated in relation to the strain energy density ψ. Thus, the Green or hyper-elastic material behavior can be defined by a pure scalar function $\psi = f(I_1, I_2, J)$, with the constitutive law:

$$\boldsymbol{\sigma} = J^{-1}\frac{d\psi}{d\mathbf{F}}\mathbf{F}^T \tag{5}$$

However, for this assumption isotropy is presumed, meaning that not every material can be modeled by this constitutive law. In order to consider anisotropic effects as well, the constitutive law is expressed by using the geometric linearized form of the right Cauchy-Green tensor.

$$\boldsymbol{\varepsilon} = \frac{1}{2}(\mathbf{F}^T\mathbf{F} - \mathbf{1}) \tag{6}$$

This modified tensor can be used to derive a simplified relation between stress and deformation. In analogy to (4) an equalization of $f(F) \sim f(\varepsilon)$ leads to Hooke's law in continuum mechanics

$$\boldsymbol{\sigma} = \frac{1}{2}(\lambda tr\boldsymbol{\varepsilon}) + 2G\boldsymbol{\varepsilon} \tag{7}$$

with the assumption $\sigma(\varepsilon = 0) = 0$ and the Lame's constants $\lambda = E\nu/((1+\nu)(1-2\nu))$ and $G = E/(2(1+\nu))$ [12].

The physical validity of this equation expires with increasing deformation due to the linearization and does not describe the stress decrease of real material. However, in practical (5) and (7) only differ for deformations higher than 5 % strain.

The vectorization of the stress and strain tensors by use of the Voigt notation leads to the well-known matrix equation:

$$\begin{bmatrix}\sigma_{11}\\ \sigma_{22}\\ \sigma_{33}\\ \sigma_{12}\\ \sigma_{13}\\ \sigma_{23}\end{bmatrix} = \frac{E}{(1+\nu)(1-2\nu)}\begin{bmatrix}1-\nu & \nu & \nu & 0 & 0 & 0\\ \nu & 1-\nu & \nu & 0 & 0 & 0\\ \nu & \nu & 1-\nu & 0 & 0 & 0\\ 0 & 0 & 0 & \frac{1-2\nu}{2} & 0 & 0\\ 0 & 0 & 0 & 0 & \frac{1-2\nu}{2} & 0\\ 0 & 0 & 0 & 0 & 0 & \frac{1-2\nu}{2}\end{bmatrix}\begin{bmatrix}\varepsilon_{11}\\ \varepsilon_{22}\\ \varepsilon_{33}\\ \varepsilon_{12}\\ \varepsilon_{13}\\ \varepsilon_{23}\end{bmatrix} \quad (8)$$

This linearized equation still describes pure isotropic material behavior. It can be converted into the generalized Hooke's law by a phenomenological motivated consideration, so that in principle all 36 coefficients can be chosen independently.

$$\begin{bmatrix}\sigma_{11}\\ \sigma_{22}\\ \sigma_{33}\\ \sigma_{12}\\ \sigma_{13}\\ \sigma_{23}\end{bmatrix} = \begin{bmatrix}C_{11} & C_{12} & C_{13} & C_{14} & C_{15} & C_{16}\\ C_{21} & C_{22} & C_{23} & C_{24} & C_{25} & C_{26}\\ C_{31} & C_{32} & C_{33} & C_{34} & C_{35} & C_{36}\\ C_{41} & C_{42} & C_{43} & C_{44} & C_{45} & C_{46}\\ C_{51} & C_{52} & C_{53} & C_{54} & C_{55} & C_{56}\\ C_{61} & C_{62} & C_{63} & C_{64} & C_{65} & C_{66}\end{bmatrix}\begin{bmatrix}\varepsilon_{11}\\ \varepsilon_{22}\\ \varepsilon_{33}\\ \varepsilon_{12}\\ \varepsilon_{13}\\ \varepsilon_{23}\end{bmatrix} \quad (9)$$

However, both the strain tensor $\boldsymbol{\varepsilon}$ and the stress tensor $\boldsymbol{\sigma}$ are symmetric, so the stiffness matrix $\boldsymbol{C}$ and the respective compliance matrix $\boldsymbol{N} = \boldsymbol{C}^{-1}$ have to be symmetric as well. Consequently, only 21 independent coefficients remain, which have to provide a positive determinant.

One should consider that this kind of anisotropic modeling is only valid for homogeneous bodies. However, anisotropic behavior in general is caused by inhomogeneity, so this is a crude assumption with very limited validity. For example, a structure causing momentums cannot be homogenized with the constitutive law (9).

However, the estimated stiffness is strongly influenced by the numerical process even for suitable structures.

2.3 Homogenization Approach

In general an analogical homogeneous constitutive law for the underlying inhomogeneous microstructure should fulfill the Hill condition [13].

$$<\boldsymbol{\sigma}><\boldsymbol{\varepsilon}>=<\boldsymbol{\sigma}\boldsymbol{\varepsilon}> \quad (10)$$

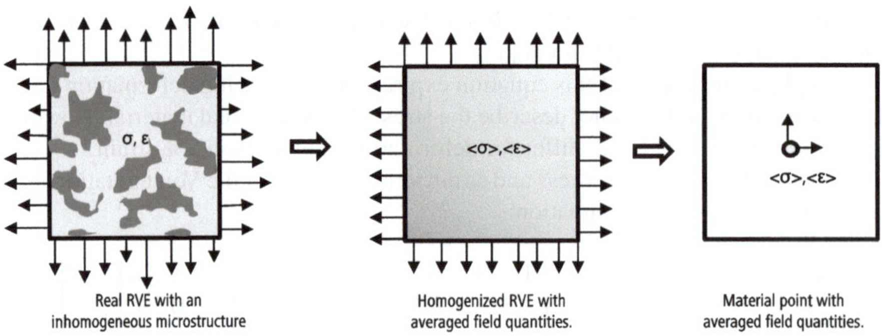

Fig. 3 Sequentially homogenization of a heterogeneous microstructure

This condition states that the average strain and stress of the heterogeneous RVE correlate with the quantities of an analogous homogeneous RVE. Strictly speaking, a perfect homogenized equation (9) of the inhomogeneous microstructure is found if this condition can be fulfilled. This equation is valid for both the analog homogeneous RVE and an arbitrary material point within the continuum. This so-called sequentially homogenization procedure is illustrated in Fig. 3.

The boundary value problem is solved by the finite element method (FEM). Any desired structures can be modelled by this procedure. Thereby, the differential equation (1) is converted to its weak form by use of the variational principle.

$$\int_{\Omega} (\sigma_{ij} n_j - t_i)\delta u_i d\Omega - \int_{V} \sigma_{ij,j}\delta u_i dV = 0 \tag{11}$$

This equation is numerically solved by discretization [14]. The field quantities are calculated discretely at the nodes of the elements and converge towards the real solution with increasing mesh refinement.

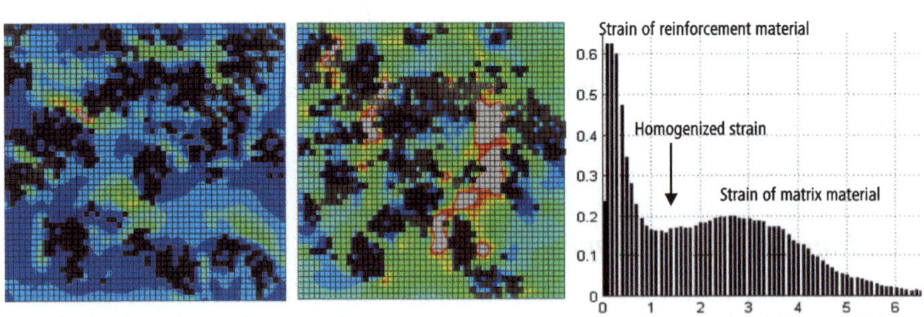

Fig. 4 *Left* Inhomogeneous strain distribution in cross sections. *Right* normalized strain distribution over the whole RVE

Figure 4 exemplary shows the inhomogeneous strain distribution in two cross sections. Additionally, the normalized distribution of all strains is plotted. Two peaks arise in the distribution, that represent the different strain levels of both materials. The macroscopic obtained strain, however, lays in between.

2.4 Window Size and Boundary Conditions

Three boundary conditions in terms of displacement or stress can be defined on each surface of the RVE, compare Figs. 5 and 6.

The Hill condition demands a compatibility of strain and (!) stress, which would require a definition of six boundary conditions on each surface, meaning that three strain and three stress components have to be applied. Consequently, the Hill condition cannot be completely fulfilled for boundary value problems like the FEM. The obligatory chosen conditions have a strong influence on the stiffness estimation. The structure is estimated too stiff, if pure displacement conditions are applied. Contrary to this, the structure is estimated too soft, if pure stress conditions are applied.

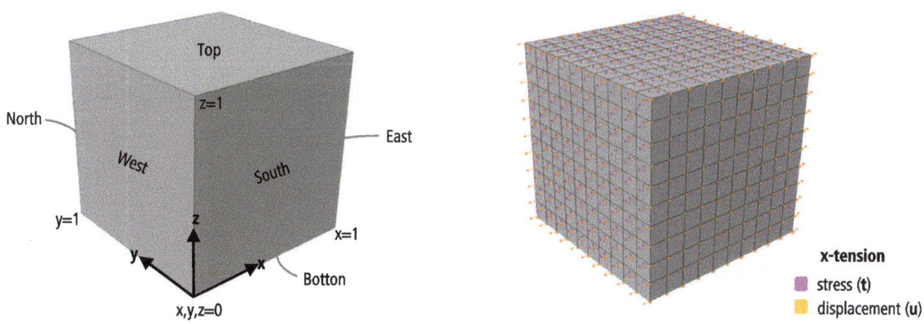

Fig. 5 *Left* Definitions of RVE surfaces. *Right* FEM visualization

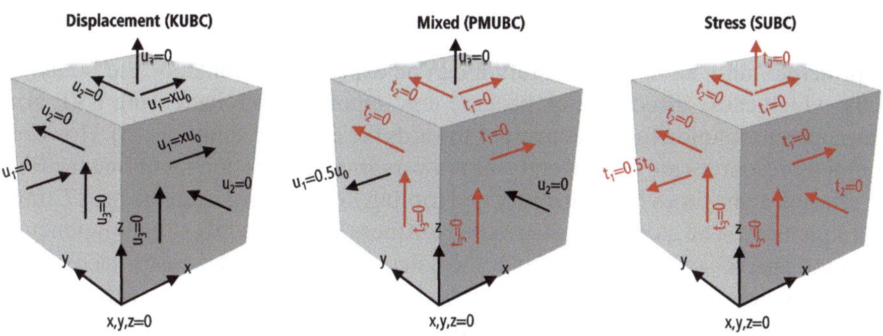

Fig. 6 Boundary condition sets. *Left* Pure kinematic constraints. *Middle* Mixed constraints proposed by Pahr and Zysset [8]. *Right* Pure stress constraints

Table 1 KUBC: Kinematic uniform boundary conditions (3 kinematic constraints on each surface)

Load case	Top	Bottom	East	West	North	South
x-tension	$u_1 = x \cdot u_0$	$u_1 = x \cdot u_0$	$u_1 = u_0$	$u_1 = 0$	$u_1 = x \cdot u_0$	$u_1 = x \cdot u_0$
	$u_2 = 0$	$u_2 = 0$	$u_2 = 0$	$u_2 = 0$	$u_2 = 0$	$u_2 = 0$
	$u_3 = 0$	$u_3 = 0$	$u_3 = 0$	$u_3 = 0$	$u_3 = 0$	$u_3 = 0$
xy-shear	$u_1 = 0$	$u_1 = 0$	$u_1 = 0$	$u_1 = 0$	$u_1 = 0$	$u_1 = 0$
	$u_2 = x \cdot u_0$	$u_2 = x \cdot u_0$	$u_2 = u_0$	$u_2 = 0$	$u_2 = x \cdot u_0$	$u_2 = x \cdot u_0$
	$u_3 = 0$	$u_3 = 0$	$u_3 = 0$	$u_3 = 0$	$u_3 = 0$	$u_3 = 0$

Table 2 PMUBC: Periodic mixed uniform boundary conditions (combination of kinematic and stress constraints)

Load case	Top	Bottom	East	West	North	South
x-tension	$t_1 = 0$	$t_1 = 0$	$u_1 = u_0/2$	$u_1 = -u_0/2$	$t_1 = 0$	$t_1 = 0$
	$t_2 = 0$	$t_2 = 0$	$t_2 = 0$	$t_2 = 0$	$u_2 = 0$	$u_2 = 0$
	$u_3 = 0$	$u_3 = 0$	$t_3 = 0$	$t_3 = 0$	$t_3 = 0$	$t_3 = 0$
xy-shear	$t_1 = 0$	$t_1 = 0$	$t_1 = 0$	$t_1 = 0$	$u_1 = u_0/2$	$u_1 = -u_0/2$
	$t_2 = 0$	$t_2 = 0$	$u_2 = u_0/2$	$u_2 = -u_0/2$	$t_2 = 0$	$t_2 = 0$
	$u_3 = 0$	$u_3 = 0$	$u_3 = 0$	$u_3 = 0$	$u_3 = 0$	$u_3 = 0$

Table 3 SUBC: Stress uniform boundary conditions (3 stress constraints on each surface)

Load case	Top	Bottom	East	West	North	South
x-tension	$t_1 = 0$	$t_1 = 0$	$t_1 = t_0/2$	$t_1 = -t_0/2$	$t_1 = 0$	$t_1 = 0$
	$t_2 = 0$	$t_2 = 0$	$t_2 = 0$	$t_2 = 0$	$t_2 = 0$	$t_2 = 0$
	$t_3 = 0$	$t_3 = 0$	$t_3 = 0$	$t_3 = 0$	$t_3 = 0$	$t_3 = 0$
xy-shear	$t_1 = 0$	$t_1 = 0$	$t_1 = 0$	$t_1 = 0$	$t_1 = t_0/2$	$t_1 = -t_0/2$
	$t_2 = 0$	$t_2 = 0$	$t_2 = t_0/2$	$t_2 = -t_0/2$	$t_2 = 0$	$t_2 = 0$
	$t_3 = 0$	$t_3 = 0$	$t_3 = 0$	$t_3 = 0$	$t_3 = 0$	$t_3 = 0$

In principle, only an apparent solution can be expected. Fortunately, the boundary cells, which cause discontinuity, become less important with increasing RVE size. The apparent solution finally converges towards the effective solution [15]. However, computer resources are limited and size is a very important factor. Doubling the RVE size means eight-times more memory and tripling the RVE size even means 27-times more memory. The stochastic distributed stiffness further reduces the efficiency of common sparse-solver. For example, the calculation of six load cases needs about 2 h on one core and 25 min on six cores (each Xeon E7-4830) for a cube with 50 elements per edge. A simulation with 100 elements per edge runs 27 h on 8 cores and needs 100 GB memory instead of 6.4 GB for 50 elements per edge.

The required RVE size can be optimized by using a RVE with geometric symmetries and periodic or mixed boundary conditions. Periodic boundary conditions should consequently have a periodic structure. In most cases this implies a complex generating of such RVE. Stochastic structures can be mirrored at three surfaces for instance. Disadvantageously, this increases the element size 8-times and enforces orthotropic behavior additionally. However, orthotropic RVEs and periodic boundary conditions lead to the effective values directly. It is observed, that for stochastic structures periodic boundary conditions show better convergence as well. Good results can be obtained using this technique, although the continuity of stress and strain is violated in the anisotropic case. PAHR and ZYSSET [8] proposed a set of mixed boundary conditions with quite similar properties to periodic conditions in combination with mirrored RVEs. So, it can be argued, that this set leads to a similar average value as real periodic boundary conditions, but with less elaboration in design. Unfortunately, no additional experiences and best practice advices concerning convergence of boundary conditions in combination with RVE size exist in the literature expect of KANIT [9, 10]. So convergence studies are an important first step. Uniform displacement conditions (KUBC) and uniform stress conditions (SUBC) are regarded additionally and compared with the PUMBC-boundary conditions of PAHR und ZYSSET. Three tension and three shear load cases are performed in total. Tables 1–3 list all sets of boundary conditions for one tension and one shear load case, respectively. The remaining four cases are defined analogical. A special feature is used for the KUBC set. Here, the constraints are continuously distributed over the surface by the variables x, y, and z.

2.5 Homogenized Anisotropy

The homogenized anisotropic stiffness matrix can be determined by inserting the average strain $<\boldsymbol{\varepsilon}>$ and stress $<\boldsymbol{\sigma}>$ into Eq. (9). Six linear independent load cases have to be simulated to specify the generalized stiffness matrix completely. This leads to 36 equations in total. Labeling the different load cases with n $= 1, 2, 3, 4, 5, 6$ leads to the following generalized form.

$$\begin{bmatrix} \langle \sigma_{11}^n \rangle \\ \langle \sigma_{22}^n \rangle \\ \langle \sigma_{33}^n \rangle \\ \langle \sigma_{12}^n \rangle \\ \langle \sigma_{13}^n \rangle \\ \langle \sigma_{23}^n \rangle \end{bmatrix} = \begin{bmatrix} C_{11} & C_{12} & C_{13} & C_{14} & C_{15} & C_{16} \\ C_{21} & C_{22} & C_{23} & C_{24} & C_{25} & C_{26} \\ C_{31} & C_{32} & C_{33} & C_{34} & C_{35} & C_{36} \\ C_{41} & C_{42} & C_{43} & C_{44} & C_{45} & C_{46} \\ C_{51} & C_{52} & C_{53} & C_{54} & C_{55} & C_{56} \\ C_{61} & C_{62} & C_{63} & C_{64} & C_{65} & C_{66} \end{bmatrix} \begin{bmatrix} \langle \varepsilon_{11}^n \rangle \\ \langle \varepsilon_{22}^n \rangle \\ \langle \varepsilon_{33}^n \rangle \\ \langle \varepsilon_{12}^n \rangle \\ \langle \varepsilon_{13}^n \rangle \\ \langle \varepsilon_{23}^n \rangle \end{bmatrix} \tag{12}$$

Considering only the first normal stress $<\boldsymbol{\sigma}_{11}^n>$ for any load case, the following inner product is obtained,

$$\langle \sigma_{11}^n \rangle = \begin{bmatrix} C_{11} & C_{12} & C_{13} & C_{14} & C_{15} & C_{16} \end{bmatrix} \begin{bmatrix} \langle \varepsilon_{11}^n \rangle \\ \langle \varepsilon_{22}^n \rangle \\ \langle \varepsilon_{33}^n \rangle \\ \langle \varepsilon_{12}^n \rangle \\ \langle \varepsilon_{13}^n \rangle \\ \langle \varepsilon_{23}^n \rangle \end{bmatrix} \tag{13}$$

which can be transformed by vector algebra into

$$\langle \sigma_{11}^n \rangle = \begin{bmatrix} \langle \varepsilon_{11}^n \rangle & \langle \varepsilon_{22}^n \rangle & \langle \varepsilon_{33}^n \rangle & \langle \varepsilon_{12}^n \rangle & \langle \varepsilon_{13}^n \rangle & \langle \varepsilon_{23}^n \rangle \end{bmatrix} \begin{bmatrix} C_{11} \\ C_{12} \\ C_{13} \\ C_{14} \\ C_{15} \\ C_{16} \end{bmatrix} \tag{14}$$

Expanding Eq. (14) to all load cases finally gives

$$\begin{bmatrix} <\sigma_{11}^1> \\ <\sigma_{11}^2> \\ <\sigma_{11}^3> \\ <\sigma_{11}^4> \\ <\sigma_{11}^5> \\ <\sigma_{11}^6> \end{bmatrix} = \begin{bmatrix} <\varepsilon_{11}^1> & <\varepsilon_{22}^1> & <\varepsilon_{33}^1> & <\varepsilon_{12}^1> & <\varepsilon_{13}^1> & <\varepsilon_{23}^1> \\ <\varepsilon_{11}^2> & <\varepsilon_{22}^2> & <\varepsilon_{33}^2> & <\varepsilon_{12}^2> & <\varepsilon_{13}^2> & <\varepsilon_{23}^2> \\ <\varepsilon_{11}^3> & <\varepsilon_{22}^3> & <\varepsilon_{33}^3> & <\varepsilon_{12}^3> & <\varepsilon_{13}^3> & <\varepsilon_{23}^3> \\ <\varepsilon_{11}^4> & <\varepsilon_{22}^4> & <\varepsilon_{33}^4> & <\varepsilon_{12}^4> & <\varepsilon_{13}^4> & <\varepsilon_{23}^4> \\ <\varepsilon_{11}^5> & <\varepsilon_{22}^5> & <\varepsilon_{33}^5> & <\varepsilon_{12}^5> & <\varepsilon_{13}^5> & <\varepsilon_{23}^5> \\ <\varepsilon_{11}^6> & <\varepsilon_{22}^6> & <\varepsilon_{33}^6> & <\varepsilon_{12}^6> & <\varepsilon_{13}^6> & <\varepsilon_{23}^6> \end{bmatrix} \begin{bmatrix} C_{11} \\ C_{12} \\ C_{13} \\ C_{14} \\ C_{15} \\ C_{16} \end{bmatrix} \tag{15}$$

The first 6 coefficients C_{11}–C_{16} of the generalized stiffness tensor can now be calculated by multiplying this equation with the inverse strain matrix $< \boldsymbol{E} >$. This enables a general notation to calculate all 36 stiffness components.

$$\begin{bmatrix} C_{i1} \\ C_{i2} \\ C_{i3} \\ C_{i4} \\ C_{i5} \\ C_{i6} \end{bmatrix} = \begin{bmatrix} <\varepsilon_{11}^1> & <\varepsilon_{22}^1> & <\varepsilon_{33}^1> & <\varepsilon_{12}^1> & <\varepsilon_{13}^1> & <\varepsilon_{23}^1> \\ <\varepsilon_{11}^2> & <\varepsilon_{22}^2> & <\varepsilon_{33}^2> & <\varepsilon_{12}^2> & <\varepsilon_{13}^2> & <\varepsilon_{23}^2> \\ <\varepsilon_{11}^3> & <\varepsilon_{22}^3> & <\varepsilon_{33}^3> & <\varepsilon_{12}^3> & <\varepsilon_{13}^3> & <\varepsilon_{23}^3> \\ <\varepsilon_{11}^4> & <\varepsilon_{22}^4> & <\varepsilon_{33}^4> & <\varepsilon_{12}^4> & <\varepsilon_{13}^4> & <\varepsilon_{23}^4> \\ <\varepsilon_{11}^5> & <\varepsilon_{22}^5> & <\varepsilon_{33}^5> & <\varepsilon_{12}^5> & <\varepsilon_{13}^5> & <\varepsilon_{23}^5> \\ <\varepsilon_{11}^6> & <\varepsilon_{22}^6> & <\varepsilon_{33}^6> & <\varepsilon_{12}^6> & <\varepsilon_{13}^6> & <\varepsilon_{23}^6> \end{bmatrix}^{-1} \begin{bmatrix} <\sigma_{ij}^1> \\ <\sigma_{ij}^2> \\ <\sigma_{ij}^3> \\ <\sigma_{ij}^4> \\ <\sigma_{ij}^5> \\ <\sigma_{ij}^6> \end{bmatrix} \tag{16}$$

However, the symmetries of the strain and stress tensors demand a symmetric stiffness tensor with only 21 components remaining. The numerical procedure could cause small deviations between the symmetric entries. Practically, they can be averaged.

2.6 Effective Isotropy

The elastic constants of a respective isotropy can be obtained with a Voigt or Reuss approximation [16].

$$\begin{aligned}
&K_{\text{Voigt}} = \tfrac{1}{9}\text{tr}\boldsymbol{C}, \text{G}_{\text{Voigt}} = \tfrac{1}{30}(3\text{tr}\boldsymbol{V} - \text{tr}\boldsymbol{C}) \\
&\text{tr}\boldsymbol{C} = C_{11} + C_{22} + C_{33} + 2(C_{12} + C_{13} + C_{23}) \\
&\text{tr}\boldsymbol{V} = C_{11} + C_{22} + C_{33} + 2(C_{44} + C_{55} + C_{66}) \\
& \\
&K_{\text{Reuss}} = \tfrac{1}{\text{tr}\boldsymbol{K}}, \text{G}_{\text{Reuss}} = \tfrac{15}{2}(3\text{tr}\boldsymbol{R} - \text{tr}\boldsymbol{K}) \\
&\text{tr}\boldsymbol{K} = N_{11} + N_{22} + N_{33} + 2(N_{12} + N_{13} + N_{23}) \\
&\text{tr}\boldsymbol{R} = N_{11} + N_{22} + N_{33} + 2(N_{44} + N_{55} + N_{66})
\end{aligned} \tag{17}$$

In theory, similar results are obtained for both approximations if the Hill condition is fulfilled. However, as described above this cannot be assumed. The advantage of these special approximations is obvious. The effective solution has to be in between those two bounds. The approximations give an ultimate upper and lower estimate of the according moduli, respectively. In practices, they can be used to calculate effective moduli by determining the arithmetic means of both estimates.

$$K_{\text{effektive}} = \frac{K_{\text{Voigt}} + K_{\text{Reuss}}}{2}, \qquad \text{G}_{\text{effektiv}} = \frac{G_{\text{Voigt}} + G_{\text{Reuss}}}{2} \tag{18}$$

The isotropic stiffness tensor can now be determined by these moduli compliant with Eq. (8). Thereby, the Young's modulus and Poisson ratio are obtained by the bulk and shear moduli as in the following

$$E = \frac{9KG}{3K + G}, \qquad \nu = \frac{3K - 2G}{2(3K + G)} \tag{19}$$

The evidence of isotropy can be checked by the Euclidean norm [8, 17, 18].

$$e = \frac{\left\| \boldsymbol{C}_{\text{isotrop}} - \boldsymbol{C}_{\text{anisotrp}} \right\|_E}{\left\| \boldsymbol{C}_{\text{isotrop}} \right\|_E} \tag{20}$$

This value gives a measure of the error and thus $e = 0$ means perfect isotropy. The determined isotropic elasticity moduli are still depending of the RVE randomness. A Monte-Carlo simulation is performed in order to obtain the desired accuracy. Thereby, the average elasticity modulus of a sufficient amount of RVEs is determined and checked for convergence.

3 Results

3.1 Monte-Carlo Simulation

Figure 7 shows the sequence of a Monte-Carlo simulation for two different RVE sizes of 20 and 50 elements. The Voigt and Reuss approximations converge separately. 50 elements per edge show a good agreement with only slight difference between both approximations.

Figure 8 shows the associated probability plots of the individual moduli. Thereby, the y-axis is scaled in a way that the values build a linear slope for a normal distribution, which is in very good agreement for these kind of RVEs. It can be seen that the graphs of the Voigt and Reuss approximations proceed closer for the larger RVE size. The variance and the anisotropy error decrease for increasing RVE size. The anisotropy error is 14.5 % for a RVE size of 20 elements per edge and 3, 4 % for a RVE size of 50 elements. The numerical extra time of large RVE can obviously be

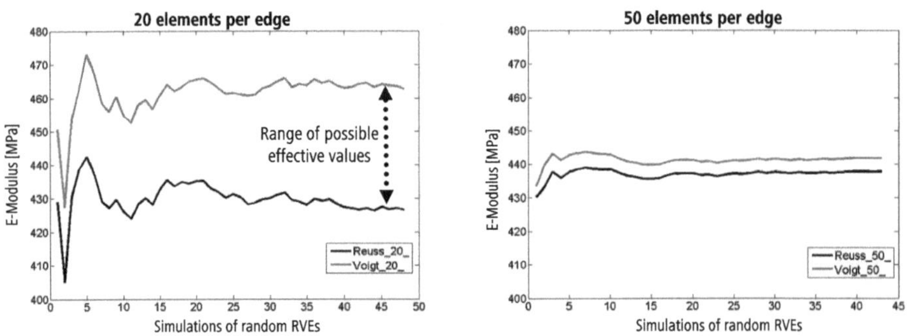

Fig. 7 Monte-Carlo simulation for RVE size of 20 (*left*) and 50 (*right*) elements per edge, respectively

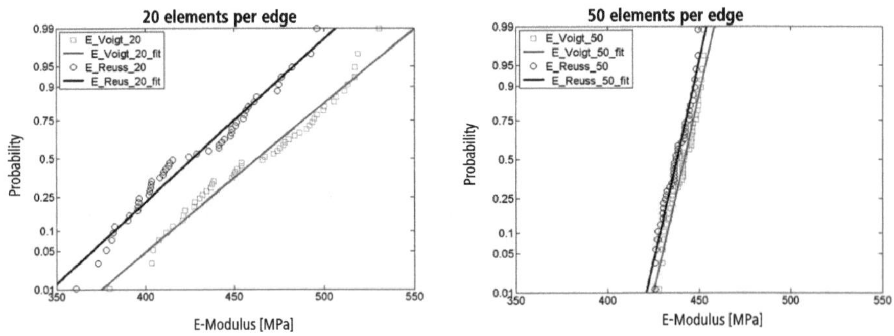

Fig. 8 Probability plot of the calculated Young's moduli for RVE size of 20 (*left*) and 50 (*right*) elements per edge, respectively

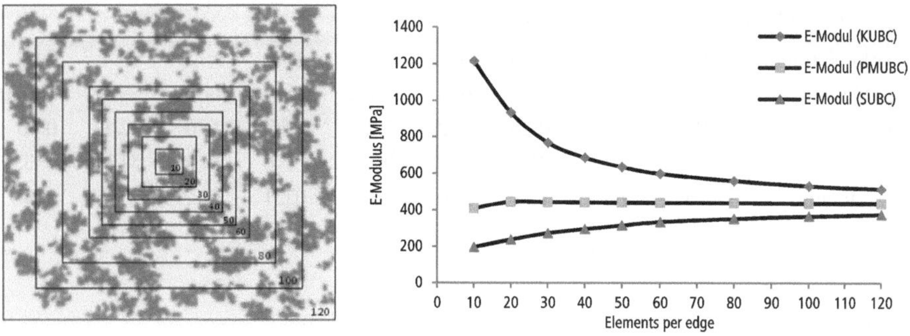

Fig. 9 Influence of boundary conditions and window size. *Left* RVE window. *Right* Dependence of the Young's modulus on number of elements per edge (window size) for varying boundary conditions

compensated by a reduced number of necessary simulations. The normal distribution allows to achieve an effective stiffness by averaging only few simulations.

3.2 RVE Size and Boundary Conditions

Figure 9 shows the convergence behavior of the three different sets of boundary conditions for a RVE with 25 % volume fraction and 0.2 % initial cells. The Young's moduli are plotted against the RVE size. Each modulus is the average of 48 simulations. Expectedly, the PMUBCs converge best. The effective modulus is nearly obtained with a RVE size of 20 elements per edge, while the KUBCs and SUBCs converge badly. These sets do not give effective moduli even for a RVE size of 120 elements per edge. As mentioned above, the KUBC estimates the structure too stiff, whereas the SUBC estimates the structure too soft. The convergence study is stopped at a RVE size of 120 elements per edge due to computational resources.

For further investigations, however, a RVE size of 50 elements per edge in combination with PUMBCs is chosen and the isotropic elasticity moduli of 48 random simulations are averaged, without rechecking for convergence.

3.3 Analysis of Stochastic Microstructures

Tables 4–6 and Fig. 10 present the results of three structure types with varying percentage amount of initial cells (0.00625, 0.2 and 6.4 %) for different volume fractions. Thereby, the respective moduli are fitted with the following empirical equation.

Table 4 Calculated moduli for 0.00625 % initial cells

Volume fraction	0.00	0.12	0.25	0.37	0.5	0.62	0.75	0.87	1.00
E-Modulus	100	169	335	688	1402	2544	4485	6903	10000
ν	0.30	0.27	0.25	0.25	0.24	0.25	0.26	0.28	0.30
K-Modulus	83	125	227	452	905	1670	3084	5124	8333
G-Modulus	38	66	133	276	565	1021	1783	2706	3846
Error		0.05	0.21	0.23	0.23	0.16	0.08	0.03	

Table 5 Calculated moduli for 0.2 % initial cells

Volume fraction	0	0.12	0.25	0.37	0.5	0.62	0.75	0.87	1
E-Modulus	100	189	442	943	1897	3179	5052	7215	10000
ν	0.30	0.27	0.24	0.23	0.23	0.24	0.26	0.28	0.30
K-Modulus	83	135	284	587	1190	2067	3503	5404	8333
G-Modulus	38	74	178	383	769	1278	2005	2824	3846
Error		0.02	0.04	0.05	0.04	0.03	0.02	0.01	

Table 6 Calculated moduli for 6.4 % initial cells

Volume fraction	0	0.12	0.25	0.37	0.5	0.62	0.75	0.87	1
E-Modulus	100	182	483	1116	2226	3608	5455	7462	10000
ν	0.30	0.26	0.23	0.23	0.23	0.25	0.26	0.28	0.30
K-Modulus	83	128	301	683	1399	2375	3838	5653	8333
G-Modulus	38	72	196	454	901	1447	2159	2915	3846
Error		0.02	0.04	0.03	0.02	0.02	0.02	0.01	

$$E = E_2 e^{bc_1} + (E_1 - E_2 e^b) c_1^b \tag{21}$$

E_1 and E_2 are the respective elasticity moduli of both materials with $E_1 > E_2$ and c_1 is the volume fraction. The shape of the curve is modeled with help of the structural parameter b.

Contrary to a volume fraction variation, three types of volume fractions are chosen to be constant (at 6, 12 and 25 %, respectively) and the percentage amount of initial cells are varied. The plots are shown in Fig. 11. The stiffness increases clearly towards a maximum with increasing dispersion. The stiffness declines again behind this maximum. This maximum, thereby, shifts with increasing volume fraction.

Equation (21) can be used to fit real measurements. Figure 12 shows the measurements of Ashmann et al. [1, 2] for bovine cancellous bone taken from femur and Rho et al. [3] for human cancellous bone taken from different bone types. The spreading of the measurements can be evolved completely, only by varying the structure parameter b. This means, the spreading can be explained by structure since both moduli are retained constant.

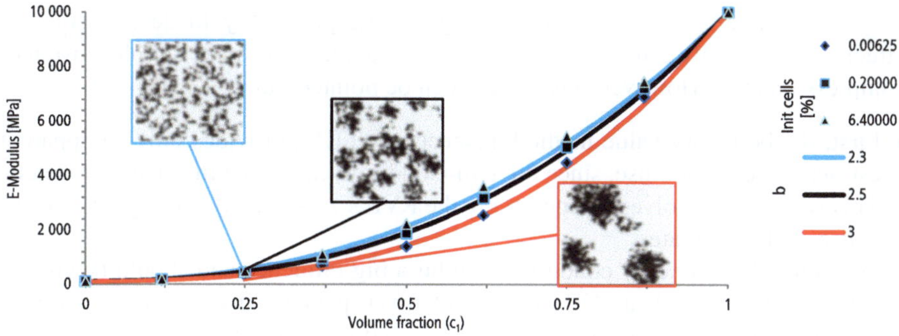

Fig. 10 E-modulus for varying volume fraction and three different reinforcement dispersions

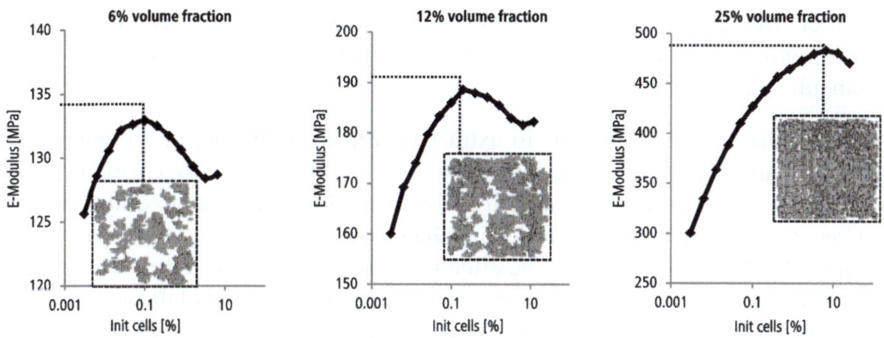

Fig. 11 E-modulus for varying initial cell percentages and three different volume fractions

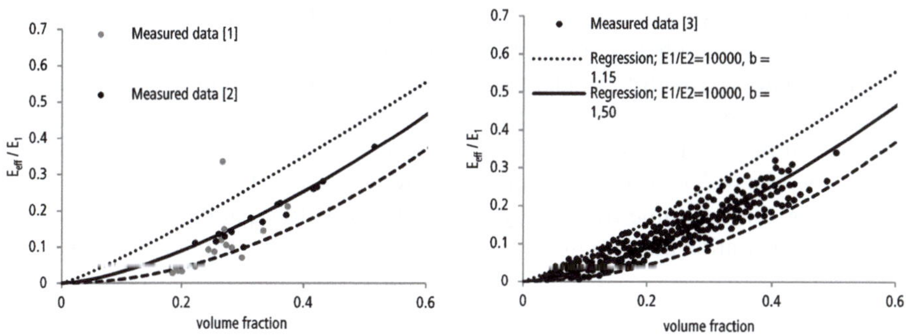

Fig. 12 Fitting of real bone measurements with Eq. (21). Data from [1–3]

4 Discussion

This work demonstrates a continuum mechanics approach to estimate effective isotropic moduli of inhomogeneous microstructures. A simple algorithm is developed to provide unlimited different RVEs with same stochastically information.

The continuum mechanics approach gives the possibility to deal anisotropic effects, which hardly can be handled with classical mechanics like simple tension or compression tests. However, three issues can be pointed out.

- First, the boundary value method, respectively FEM, provides only an apparent estimate in common use, since the Hill-Condition cannot be fulfilled completely. This issue can be solved by increasing the RVE size and by regarding the convergence of the estimate.
- Secondly, the boundary conditions exhibit a big influence and should be chosen carefully. In this study a RVE size of 50 elements per edge in combination with the PMUBC set of Zysset and Pahr established as good compromise with respect to the computational resources. Pure displacement or stress conditions are not working.
- Since each RVE represents a random window of the structure, the randomness can be averaged by using the Monte-Carlo procedure at last. It could be shown that multiple simulations show a normal distribution and simple averaging of the moduli works.

An additional scientific issue is to extract isotropic moduli out of an anisotropic stiffness matrix since there are unlimited possible combinations. This work proposes a procedure based on a Voigt and Reuss approximation. These approximations give the lower and upper bounds, representing the softest and the stiffest combinations of the moduli. While knowing that the effective moduli have to be in-between, simple averaging leads to the effective moduli in practice. Both bounds fall together for a sufficient RVE size. A RVE size of 50 elements per edge gives very close bounds.

The effective moduli are used to build an isotropic stiffness matrix, which can be compared to the original anisotropic matrix by the Euclidian error measure.

The influences of percentage amount of initial cells and volume fraction were studied. Since the number of initial cells controls the structure an empirical regression equation could be found to model this behavior. A specific number of initial cells could be determined to reach a maximal stiffness with respect to each volume fraction. The algorithm is not made to model real bone structure, however, the empirical equation works well to model real bone measurements. This model predicts different stiffness for same volume fraction by varying the structure parameter. In principle, the stiffness variation that is observed in measurements can be explained by structure only.

References

1. Ashman, R.B., Corin, J.D., Turner, C.H.: Elastic properties of cancellous bone: measurement by an ultrasonic technique. J. Biomech. **20**, 979–986 (1987)
2. Ashman, R.B., Rho, J.Y.: Elastic modulus of trabecular bone material. J. Biomech. **21**, 177–181 (1988)
3. Rho, J.Y., Ashman, R.B., Turner, C.H.: Youngs modulus of trabecular and cortical bone material: ultrasonic and microtensile measurements. J. Biomech. **26**, 111–119 (1993)

4. Rho, J.Y., Tsui, T.Y., Pharr, G.M.: Elastic properties of human cortical and trabecular lamellar bone measured by nanoindentation. Biomaterials **18**, 1325–1330 (1997)
5. Dalstra, M., Huiskes, R., Odgaard, A., van Erning, L.: Mechanical and textural properties of pelvic trabecular bone. J. Biomech. **26**, 523–535 (1993)
6. Müller, R., Rüegsegger, P.: Analysis of mechanical properties of cancellous bone under conditions of simulated bone atrophy. J. Biomech. **29**, 1053–1060 (1996)
7. Ulrich, D., van Rietbergen, B., Weinans, H., Rüegsegger, P.: Finite element analysis of trabecular bone structure: a comparison of image-based meshing techniques. J. Biomech. **31**, 1187–1192 (1998)
8. Pahr, D.H., Zysset, P.K.: Influence of boundary conditions on computed apparent elastic properties of cancellous bone. Biomech. Model. Mechanobiol. **7**, 463–476 (2008)
9. Kanit, T., Forest, S., Galliet, I., Mounoury, V., Jeulin, D.: Determination of the size of the representative volume element for random composites: statistical and numerical approach. Int. J. Solid. Struct. **40**(13–14), 3647–3679 (2003). doi:10.1016/S0020-7683(03)00143-4
10. Kanit, T., N'Guyen, F., Forest, S., Jeulin, D., Reed, M., Singleton, S.: Apparent and effective physical properties of heterogeneous materials: representativity of samples of two materials from food industry. Comput. Methods Appl. Mech. Eng. **195**(33–36), 3960–3982 (2006). doi:10.1016/j.cma.2005.07.022
11. Holzapfel, G.A.: Nonlinear Solid Mechanics. A Continuum Approach for Engineering. Wiley, Chichester, New York (2000)
12. Greve, R.: Kontinuumsmechanik. Springer, Berlin (2003)
13. Hill, R.: Elastic properties of reinforced solids: some theoretical principles. J. Mech. Phys. Solid. **11**(5), 357–372 (1963)
14. Washizu, Kyuichiro: Variational Methods in Elasticity and Plasticity, 3rd edn. Pergamon Press, Oxford, New York (1982)
15. Sab, K.: On the homogenization and the simulation of random materials. Eur. J. Mech. A/Solid **11**(5), 585–607 (1992)
16. Cowin, S.C.: Properties of the anisotropic elasticity tensor. Q. J. Mech. Appl. Math. **42**, 249–266 (1989)
17. Johnson, L.W., Riess, R.D.: Numerical Analysis, 2nd edn. Addison-Wesley Pub. Co., Reading, Mass (1982)
18. Gavazzi, A.C., Lagoudas, D.C.: On the numerical evaluation of Eshelby's tensor and its application to elastoplastic fibrous composites. Comput. Mech. **7**, 13–19 (1990)

A Gradient-Enhanced Continuum Damage Model for Residually Stressed Fibre-Reinforced Materials at Finite Strains

Tobias Waffenschmidt, César Polindara and Andreas Menzel

Abstract The modelling of damage effects in materials constitutes a major challenge in various engineering-related disciplines. However, the assumption of purely *local* continuum damage formulations may lead to ill-posed boundary value problems and—with regard to numerical methods such as the finite element method—to mesh-dependent solutions, a vanishing localised damage zone upon mesh refinement, and hence physically questionable results. In order to circumvent these deficiencies, we present a *non-local gradient-enhanced* damage model at finite strains. We additively compose the hyperelastic constitutive response at local material point level of an isotropic matrix and of an anisotropic fibre-reinforced material. The inelastic constitutive response is characterised by a scalar $[1-d]$-damage model, where we assume only the anisotropic elastic part to damage. Furthermore, we enhance the local free energy by a gradient-term. This term essentially contains the gradient of the non-local damage variable which we introduce as an additional global field variable. In order to guarantee the equivalence between the local and non-local damage variable, we incorporate a penalisation term within the free energy. Based on the principle of minimum total potential energy, we obtain a coupled system of variational equations. The associated non-linear system of equations is symmetric and can conveniently be solved by standard incremental-iterative Newton-Raphson schemes or arc-length-based solution methods. As a further key aspect, we incorporate *residual stresses* by means of a multiplicative composition of the deformation gradient. As a three-dimensional finite element example, we study the material degradation of a fibre-reinforced tube subjected to internal pressure. This highlights the mesh-objective and constitutive properties of the model and illustratively underlines the capabilities of the formulation with regard to biomechanical application such as the simulation of arteries.

Keywords Gradient-enhanced damage · Large deformations · Finite element method · Residual stresses · Anisotropic biological tissues

---

T. Waffenschmidt (✉) · C. Polindara · A. Menzel
Department of Mechanical Engineering, Institute of Mechanics, Technical University Dortmund, Leonhard-Euler-Str. 5, 44227 Dortmund, Germany
e-mail: tobias.waffenschmidt@udo.edu

A. Menzel
Division of Solid Mechanics, Lund University, P.O. Box 118, SE-22100 Lund, Sweden

T. Lenarz and P. Wriggers (eds.), *Biomedical Technology*, Lecture Notes in Applied and Computational Mechanics 74, DOI 10.1007/978-3-319-10981-7_2

1 Introduction

The physical understanding and modelling of damage and failure in materials presents a major challenge in various engineering-related disciplines. To give an example, the modelling of damage effects in anisotropic soft biological tissues is closely related to the continuous failure of fibres embedded in the ambient matrix material. Arteries, for instance, can be considered as a composite of an isotropic ground substance of elastin fibres and a highly anisotropic network of cross-linked collagen fibrils. Mechanical loading beyond the physiological loading range, e.g. caused by a surgical intervention such as balloon angioplasty, can significantly reduce the elastic properties of the artery. These softening phenomena can be attributed to the continuous overstretch and degradation of particular collagen fibres.

Based on the classical work by Kachanov [8], who associated damage effects to an area reduction of the stress-bearing region, material degradation can be modelled by means of standard continuum damage formulations, i.e. in a local sense. Up to now, a large variety of models exist where we refer the reader to classic monographs and textbooks as, e.g. Krajcinovic and Lemaitre [9], or Lemaitre [10], to name only two. However, the assumption of a purely *local* continuum damage may imply the ill-posedness of the underlying boundary value problem. With regard to numerical methods such as the finite element method, this may lead to mesh-dependent solutions, a vanishing localised damage zone upon mesh refinement, and hence physically questionable results.

In order to circumvent these deficiencies, i.e. to *regularise* the problem, several approaches have been proposed in the literature as, for instance, the concept of generalised *non-local* continua, where internal length scales are introduced into the continuum formulation, see the monograph by Eringen [4], the article by Aifantis [1] or the contributions in Eringen [3] and Rogula [13]. A non-local continuum formulation can generally be established by either introducing an integral- or a gradient-type equation.

In this contribution, we apply a non-local gradient-based damage formulation within a geometrically non-linear setting allowing for large deformations. Conceptually following Dimitrijević and Hackl [2], we enhance the local free energy by a gradient-term. This term essentially contains the gradient of the non-local damage variable which, itself, is introduced as an additional global field variable. In order to guarantee the equivalence between the local and non-local damage variable, a penalisation term is incorporated within the free energy. Based on the principle of minimum total potential energy, a coupled system of Euler-Lagrange equations, i.e. the balance of linear momentum and the balance corresponding to the non-local damage field, is obtained and solved in weak form. As a key aspect, the hyperelastic constitutive response at local material point level is governed by a highly non-linear strain energy, additively composed of an isotropic matrix material and of an anisotropic contribution related to the fibre-reinforcement. The inelastic constitutive response is represented by a scalar $[1-d]$-damage formulation, where only the anisotropic elastic part is assumed to damage. The resulting coupled, highly non-linear system of equations is symmetric and can conveniently be solved by standard incremental-iterative Newton-Raphson schemes or arc-length-based solution methods without any need for advanced and computationally expensive solution methods such as a global active-set-search as applied by Liebe et al. [11]. Furthermore, the approach proves to be robust, even at high levels of degradation, and allows to incorporate any suitable scalar-valued damage formulation.

Apart from the highly nonlinear elastic response of soft biological tissues, it is well-known that the structural design of arteries is characterised by a fibre-reinforced multi-layered composite subjected to pronounced *residual stresses*. The complex interaction of material properties with these residual stress effects and the geometry guarantees the optimal support under different blood pressures within the vessel. As a further key aspect of this contribution, we therefore incorporate residual stresses by means of a multiplicative composition of the deformation gradient.

This article is structured as follows: In Sect. 2, we summarise relevant kinematic relations for the geometrically non-linear case and the balance equations of the coupled boundary value problem in weak form. In Sect. 3, we specify the underlying constitutive equations, containing the isotropic and anisotropic non-linear elastic and gradient-enhanced free energies, as well as the continuum damage formulation. In Sect. 4, we discretise the governing weak forms by means of the finite element method resulting in a coupled non-linear system of equations. Last, in Sect. 6, we apply the model to illustrative three-dimensional inhomogeneous deformation problems. In order to show the capabilities of the approach with regard to biomechanics-related problems, we study a force-driven finite element example by means of an anisotropic artery-like tube subjected to internal pressure and under consideration of residual stresses. We conclude with a summary and future perspectives in Sect. 7.

2 Gradient Enhancement of a Continuum Damage Formulation

This section summarises the essential kinematic relations and presents the governing coupled balance equations of the boundary value problem on the basis of the principle of minimum total potential energy. The related variational form provides the basis for the finite element discretisation described in Sect. 4.

2.1 Basic Kinematics

Let $\boldsymbol{x} = \boldsymbol{\varphi}(\boldsymbol{X}, t)$ describe the deformation of the body, which transforms referential placements $\boldsymbol{X} \in \mathcal{B}_0$ to their spatial counterparts $\boldsymbol{x} \in \mathcal{B}_t$. Based on this, the deformation gradient is defined as $\boldsymbol{F} = \nabla_{\boldsymbol{X}} \boldsymbol{\varphi}$ which transforms infinitesimal referential line elements $\mathrm{d}\boldsymbol{X}$ onto their spatial counterparts $\mathrm{d}\boldsymbol{x}$. Furthermore, let $\mathrm{d}V$ and $\mathrm{d}v$ denote an infinitesimal volume element in referential and spatial setting. Accordingly, the Jacobian $J = \mathrm{d}v/\mathrm{d}V = \det(\boldsymbol{F}) > 0$ is the ratio of the deformed to the undeformed volume. Finally, let $\mathrm{d}\boldsymbol{A} = \boldsymbol{N}\,\mathrm{d}A$ and $\mathrm{d}\boldsymbol{a} = \boldsymbol{n}\,\mathrm{d}a$ define the referential and spatial area normals. Then, Nanson's formula $\boldsymbol{n}\,\mathrm{d}a = \mathrm{cof}(\boldsymbol{F}) \cdot \boldsymbol{N}\,\mathrm{d}A$ describes the transformation of infinitesimal area elements between the reference and the spatial configuration with the co-factor of $\boldsymbol{F}$ defined as $\mathrm{cof}(\boldsymbol{F}) = J\boldsymbol{F}^{-\mathrm{t}}$. Fibre-reinforcement of the material is incorporated by assuming two families of fibres to be embedded in the continuum. Their orientation is characterised by referential unit vectors $\boldsymbol{a}_{0\,i}$, $i = 1, 2$ with $||\boldsymbol{a}_{0\,i}|| = 1$.

2.2 General Gradient-Enhanced Format of the Free Energy

We assume the local free energy

$$\Psi_{\mathrm{loc}}(\boldsymbol{F}, \boldsymbol{a}_{0\,i}, \kappa) = \Psi_{\mathrm{iso}}(\boldsymbol{F}) + f_{\mathrm{d}}(\kappa)\, \Psi_{\mathrm{ani}}(\boldsymbol{F}, \boldsymbol{a}_{0\,i}) \tag{1}$$

to account for anisotropic non-linear elastic material response under the influence of scalar damage. Basically, we additively compose the effective free energy of the undamaged material of an isotropic contribution Ψ_{iso} representing the ground substance and of an anisotropic contribution Ψ_{ani} associated with i fibre families. We assume only the anisotropic part to be affected by the damage, whereas the isotropic matrix material remains elastic. In Eq. (1), $\kappa \in [0, \infty)$ is a scalar internal damage variable, characterising a material stiffness loss of the fibres, while $f_{\mathrm{d}}(\kappa) = 1 - d \in (0, 1]$ represents an appropriate damage function that is at least twice differentiable and satisfies $f_{\mathrm{d}}(0) = 1$ and $f_{\mathrm{d}}(\kappa \to \infty) \to 0$. This ensures purely elastic behaviour of the undamaged material and complete loss of the related material stiffness for $\kappa \to \infty$. Conceptually following the approach by Dimitrijević and Hackl [2], we introduce a gradient-enhanced non-local free energy Ψ_{nloc} as

$$\Psi_{\mathrm{nloc}}(\phi, \nabla_X \phi, \kappa; \boldsymbol{F}) = \Psi_{\mathrm{grd}}(\nabla_X \phi; \boldsymbol{F}) + \Psi_{\mathrm{plty}}(\phi, \kappa)\,. \tag{2}$$

Here, Ψ_{grd} contains the referential gradient of the non-local damage field variable ϕ while Ψ_{plty} incorporates a penalisation term which links the non-local damage variable ϕ to the local damage variable κ. Consequently, we obtain an enhanced free energy as

$$\Psi_{\mathrm{int}}(\boldsymbol{F}, \phi, \nabla_X \phi, \boldsymbol{a}_{0\,i}, \kappa) = \Psi_{\mathrm{loc}}(\boldsymbol{F}, \boldsymbol{a}_{0\,i}, \kappa) + \Psi_{\mathrm{nloc}}(\phi, \nabla_X \phi, \kappa; \boldsymbol{F})\,. \tag{3}$$

Provided that the external load can be derived from a potential, we can specify the local external energy function as $\Psi_{\mathrm{ext}}(\boldsymbol{\varphi})$. In summary, the total potential energy function is additively composed of the internal and external contribution so that its local form reads

$$\Psi(\boldsymbol{\varphi}, \boldsymbol{F}, \phi, \nabla_X\phi, \boldsymbol{a}_{0\,i}, \kappa) = \Psi_{\text{int}}(\boldsymbol{F}, \phi, \nabla_X\phi, \boldsymbol{a}_{0\,i}, \kappa) + \Psi_{\text{ext}}(\boldsymbol{\varphi}) \,. \tag{4}$$

2.3 Total Potential Energy

The total potential energy of a system additively combines the internal contribution Π_{int}, reflecting the action of internal forces, and an external contribution $\Pi_{\text{ext}} = \Pi_{\text{vol}} + \Pi_{\text{sur}}$ due to volume and surface forces, i.e.

$$\Pi(\boldsymbol{\varphi}, \boldsymbol{F}, \phi, \nabla_X\phi; \boldsymbol{a}_{0\,i}, \kappa) = \Pi_{\text{int}} + \Pi_{\text{ext}} \,. \tag{5}$$

The internal energy contribution can be written as

$$\Pi_{\text{int}}(\boldsymbol{F}, \phi, \nabla_X\phi; \boldsymbol{a}_{0\,i}, \kappa) = \int_{\mathcal{B}_0} \Psi_{\text{int}} \, \mathrm{d}V \,. \tag{6}$$

while the external contributions, assuming 'dead' loads, are provided by

$$\Pi_{\text{vol}}(\boldsymbol{\varphi}) = \int_{\mathcal{B}_0} \Psi_{\text{vol}} \, \mathrm{d}V = -\int_{\mathcal{B}_0} \bar{\boldsymbol{B}} \cdot \boldsymbol{\varphi} \, \mathrm{d}V \,, \tag{7}$$

$$\Pi_{\text{sur}}(\boldsymbol{\varphi}) = \int_{\partial\mathcal{B}_0} \Psi_{\text{sur}} \, \mathrm{d}A = -\int_{\partial\mathcal{B}_0} \bar{\boldsymbol{T}} \cdot \boldsymbol{\varphi} \, \mathrm{d}A \,, \tag{8}$$

where $\bar{\boldsymbol{B}}$ denotes the body force vector per unit reference volume and $\bar{\boldsymbol{T}}$ characterises the traction vector per unit reference surface area. In this regard, see, for instance, Waffenschmidt and Menzel [15] where a double-layered thick-walled cylindrical tube subjected to internal pressure is analysed on the basis of a total potential.

2.4 Variational Form

The boundary value problem is governed by the principle of minimum potential energy

$$\min_{\varphi, \phi} \Pi(\boldsymbol{\varphi}, \boldsymbol{F}, \phi, \nabla_X\phi; \boldsymbol{a}_{0\,i}, \kappa) \,, \tag{9}$$

which requires the first variation of the total potential energy with respect to $\boldsymbol{\varphi}$ and ϕ to vanish, i.e.

$$\delta_{\boldsymbol{\varphi}} \Pi(\boldsymbol{\varphi}, \boldsymbol{F}, \phi, \nabla_{\boldsymbol{X}}\phi; \boldsymbol{a}_{0\,i}, \kappa) \doteq \boldsymbol{0} \quad \forall \;\; \delta\boldsymbol{\varphi}\,, \tag{10}$$

$$\delta_{\phi} \Pi(\boldsymbol{\varphi}, \boldsymbol{F}, \phi, \nabla_{\boldsymbol{X}}\phi; \boldsymbol{a}_{0\,i}, \kappa) \doteq 0 \quad \forall \;\; \delta\phi\,. \tag{11}$$

Taking into account that $\delta_{\boldsymbol{\varphi}} W_{\text{int}} = \delta_{\boldsymbol{\varphi}} \Pi_{\text{int}}$ and $\delta_{\boldsymbol{\varphi}} W_{\text{ext}} = -\delta_{\boldsymbol{\varphi}} \Pi_{\text{ext}}$, we obtain a coupled system of variational equations, i.e.

$$\delta_{\boldsymbol{\varphi}} W = \delta_{\boldsymbol{\varphi}} W_{\text{int}} - \delta_{\boldsymbol{\varphi}} W_{\text{ext}} = \boldsymbol{0} \quad \forall \;\; \delta\boldsymbol{\varphi}\,, \tag{12}$$

$$\delta_{\phi} W = \delta_{\phi} W_{\text{int}} - \delta_{\phi} W_{\text{ext}} = 0 \quad \forall \;\; \delta\phi\,, \tag{13}$$

where the internal and external contributions are given in spatial form as

$$\delta_{\boldsymbol{\varphi}} W_{\text{int}} = \int_{\mathcal{B}_t} \boldsymbol{\sigma} : \nabla_{\boldsymbol{x}}\delta\boldsymbol{\varphi}\,\text{d}v\,, \qquad \delta_{\boldsymbol{\varphi}} W_{\text{ext}} = \int_{\mathcal{B}_t} \bar{\boldsymbol{b}} \cdot \delta\boldsymbol{\varphi}\,\text{d}v + \int_{\partial\mathcal{B}_t} \bar{\boldsymbol{t}} \cdot \delta\boldsymbol{\varphi}\,\text{d}a\,, \tag{14}$$

$$\delta_{\phi} W_{\text{int}} = \int_{\mathcal{B}_t} \boldsymbol{y} \cdot \nabla_{\boldsymbol{x}}\delta\phi\,\text{d}v\,, \qquad \delta_{\phi} W_{\text{ext}} = \int_{\mathcal{B}_t} y\,\delta\phi\,\text{d}v\,. \tag{15}$$

Here the Cauchy stress $\boldsymbol{\sigma}$ and the vectorial damage quantity $\boldsymbol{y}$ are related to *flux* terms, whereas the body force $\bar{\boldsymbol{b}}$ and the scalar damage quantity y are associated to *source* terms. They are defined as

$$\boldsymbol{\sigma} = \partial_{\boldsymbol{F}}\Psi \cdot \text{cof}(\boldsymbol{F}^{-1})\,, \qquad \bar{\boldsymbol{b}} = -J^{-1}\partial_{\boldsymbol{\varphi}}\Psi_{\text{vol}}\,, \tag{16}$$

$$\boldsymbol{y} = \partial_{\nabla_{\boldsymbol{X}}\phi}\Psi \cdot \text{cof}(\boldsymbol{F}^{-1})\,, \qquad y = -J^{-1}\partial_{\phi}\Psi\,. \tag{17}$$

Relations (14) and (15) provide the basis for the finite element discretisation in Sect. 4.

3 Constitutive Relations

In this section, we first review the hyperelastic constitutive equations adopted on the basis of an isotropic neo-Hookean relation and an anisotropic exponential part. These relations characterise the *elastic* anisotropic response of the fibre-reinforced material. Secondly, we specify the gradient-enhanced, non-local contribution to the free energy, followed by the continuum damage formulation.

3.1 Hyperelastic Part of the Free Energy

From Sect. 2.2, we recall the local free energy density Ψ_{loc}, Eq. (1), to be additively composed of an isotropic part Ψ_{iso}, representing the isotropic matrix material, and of an anisotropic part Ψ_{ani}, representing the individual families of fibres. In the following, we assume the isotropic part to be specified by a compressible neo-Hookean format

$$\Psi_{\text{iso}} = \frac{\mu_e}{2}[I_1 - 3] - \mu_e \ln(J) + \frac{\lambda_e}{2}[\ln(J)]^2, \tag{18}$$

with $I_1 = \boldsymbol{F} : \boldsymbol{F}$ denoting the first invariant. The elastic constants are represented by the Lamé-parameters μ_e and $\lambda_e = \kappa_e - 2/3\,\mu_e$ in terms of the shear modulus μ_e and the bulk modulus κ_e. The anisotropic contribution of the local free energy (1) is based on an orthotropic exponential model with two families of fibres including fibre dispersion according to Gasser et al. [5] or Menzel et al. [12], i.e.

$$\Psi_{\text{ani}} = \frac{k_1}{2\,k_2}\sum_{i=1}^{N}\left[\exp\left(k_2\,\langle E_i\rangle^2\right) - 1\right], \tag{19}$$

with the strain-like quantity $E_i = \varkappa I_1 + [1 - 3\varkappa]\,I_{4\,i} - 1$ and the invariant $I_{4\,i} = \boldsymbol{a}_{0\,i} \cdot \boldsymbol{F}^{\text{t}} \cdot \boldsymbol{F} \cdot \boldsymbol{a}_{0\,i}$ for $N = 2$ fibre families. The term $\langle E_i\rangle$, where $\langle\bullet\rangle = [|\bullet| + \bullet]/2$ is the Macaulay bracket, reflects the assumption that fibres can support tension only. Consequently, $\Psi_{\text{ani}} > 0$ only if the fibre-related strain is positive, i.e. $E_i > 0$. Fibre dispersion is introduced by means of the parameter $\varkappa \in [0,\ 1/3]$, where $\varkappa = 0$ corresponds to no dispersion, i.e. transverse isotropy, and where $\varkappa = 1/3$ renders an isotropic distribution. Table 1 summarises the structural and elastic material quantities included in constitutive Eqs. (18) and (19) together with their units. It is important to note that the fibre orientations may be defined arbitrarily, but the present formulation uses only *one* non-local damage variable so that both fibre families undergo identical degradation. This is physically meaningful as long as both families of fibers possess one and the same stretch history, otherwise a second non-local damage variable should be included in the formulation.

Table 1 Constitutive parameters as used in Sects. 3.1–3.3

Type	Symbol	Description	Unit
Structural	$\boldsymbol{a}_{0\,i}$	Fibre orientation vectors	–
	$\varkappa$	Dispersion parameter	–
Elastic	μ_e	Shear modulus	kPa
	κ_e	Bulk modulus	kPa
	k_1	Elastic constant	kPa
	k_2	Elastic constant	–
Regularisation	c_d	Degree of regularisation	$\text{kPa}^{-1}\text{mm}^2$
	β_d	Penalty parameter	kPa^{-1}
Damage	η_d	Saturation parameter	kPa^{-1}
	κ_d	Damage threshold	kPa

3.2 Gradient-Enhanced Part of the Free Energy

According to Eq. (2), we assume the non-local part of the free energy to be additively composed of a gradient-related term Ψ_{grd} and of a penalty term Ψ_{plty} and specify these terms as

$$\Psi_{\text{grd}}(\nabla_{\boldsymbol{X}}\phi;\, \boldsymbol{F}) = \frac{c_{\text{d}}}{2}\,||\nabla_{\boldsymbol{x}}\phi\,||^2 \quad \text{and} \quad \Psi_{\text{plty}}(\phi,\kappa) = \frac{\beta_{\text{d}}}{2}\,[\phi-\kappa]^2\,. \tag{20}$$

The energy-related penalty parameter β_{d} approximately enforces the local damage field κ and the non-local field ϕ to coincide. Furthermore, the gradient parameter c_{d} controls the quasi-non-local character of the formulation and characterises the degree of gradient regularisation: $c_{\text{d}} = 0$ results in a local model, while $c_{\text{d}} > 0$ leads to the regularised gradient-enhanced model. The damage-related parameters included in constitutive Eq. (20) together with their units are summarised in Table 1.

3.3 Gradient-Enhanced Damage Model

In order to obtain the stress-like thermodynamic forces driving the local dissipative damage process, we follow the standard Coleman-Noll procedure. Differentiation of the general format of the free energy (3) with respect to time and application of the Clausius-Planck inequality yields, amongst others, a contribution including the thermodynamic force $q = q_{\text{loc}} + q_{\text{nloc}} = -\,\partial_d\Psi$ conjugate to the damage variable d, i.e.

$$q_{\text{loc}} = \Psi_{\text{ani}} \quad \text{and} \quad q_{\text{nloc}} = \beta_{\text{d}}\,[\phi-\kappa]\,\partial_d\kappa\,. \tag{21}$$

Furthermore, we adopt the damage condition

$$\Phi_{\text{d}} = q - \kappa \le 0 \tag{22}$$

where $\Phi_{\text{d}} < 0$ refers to the purely elastic loading and $\Phi_{\text{d}} = 0$ to damage evolution. Based on the postulate of maximum dissipation, we construct a constrained optimisation problem involving the Lagrange multiplier λ. This results in the following associated evolution equation for the damage variable

$$\dot{\kappa} = \lambda\,\frac{\partial\Phi_{\text{d}}}{\partial q} = \lambda \quad \text{with} \quad \kappa|_{t=0} = \kappa_{\text{d}}\,. \tag{23}$$

where initiation and termination of damage are governed by the Karush-Kuhn-Tucker complementary conditions

$$\lambda \ge 0\,, \quad \Phi_{\text{d}} \le 0\,, \quad \lambda\,\Phi_{\text{d}} = 0\,. \tag{24}$$

We assume an exponential behaviour for the damage function

$$f_\mathrm{d}(\kappa) = 1 - d = \exp(\eta_\mathrm{d}\,[\kappa_\mathrm{d} - \kappa])\,, \tag{25}$$

with $\eta_\mathrm{d} > 0$ so that $\partial_d \kappa = [\eta_\mathrm{d}\, f_\mathrm{d}(\kappa)]^{-1} > 0$ and introduce an initial damage threshold κ_d, which must be exceeded in order to activate damage evolution. Furthermore, we include a saturation parameter η_d. It becomes apparent that larger values of η_d accelerate the damage process, whereas larger values of κ_d lead to a delay of the damage initiation. Note, that for the limiting case $\kappa_\mathrm{d} = 0$, damage is initiated from the very beginning of the loading process, whereas damage does not evolve for $\kappa_\mathrm{d} \to \infty$.

4 Finite Element Discretisation

This section deals with the spatial finite element discretisation of the underlying coupled system of non-linear equations. This includes a combination of tri-quadratic serendipity interpolation functions with respect to the displacement field, and tri-linear interpolation functions with respect to the non-local damage field variable where we outline an efficient and compact FE-implementation using a common Voigt-notation-based vector-matrix-format.

4.1 Discretisation

We discretise the domain $\mathcal{B}_0$ by n_el finite elements, so that $\mathcal{B}_0 \approx \mathcal{B}_0^\mathrm{h} = \bigcup_{e=1}^{n_\mathrm{el}} \mathcal{B}_{0\,e}$, where every finite element $\mathcal{B}_{0\,e}$ is characterised by n_en^φ placement-nodes and n_en^ϕ non-local-damage-nodes. According to the isoparametric concept, we interpolate the field variables $\boldsymbol{\varphi}$ as well as the geometry $\boldsymbol{X}$ by the same shape functions N^φ

$$\boldsymbol{X} \approx \boldsymbol{X}^\mathrm{h} = \boldsymbol{N}^\varphi \cdot \boldsymbol{X}_e, \qquad\qquad \boldsymbol{\varphi} \approx \boldsymbol{\varphi}^\mathrm{h} = \boldsymbol{N}^\varphi \cdot \boldsymbol{\varphi}_e, \tag{26}$$

and transform them to a hexahedral reference element with natural coordinates $\boldsymbol{\xi} := \{\xi, \eta, \zeta\} \in \bar{\mathcal{B}}$, where $\bar{\mathcal{B}} := \{\boldsymbol{\xi} \in \mathbb{R}^3 | -1 \le \chi \le +1;\ \chi = \xi, \eta, \zeta\}$ denotes the reference domain. In the present context, the number of displacement-nodes and non-local-damage-nodes per element—and consequently the related shape functions—do not necessarily have to coincide, i.e. $n_\mathrm{en}^\varphi \neq n_\mathrm{en}^\phi$ and $N^\varphi \neq N^\phi$. We approximate the associated field variables, i.e. the placement $\boldsymbol{\varphi}$ and the non-local damage variable ϕ by means of the product of shape functions $N^\alpha(\boldsymbol{\xi})$, $\alpha = \varphi, \phi$ and discrete element degrees of freedom $\boldsymbol{\varphi}_e$ and $\boldsymbol{\phi}_e$, i.e.

$$\boldsymbol{\varphi}^\mathrm{h} = \boldsymbol{N}^\varphi \cdot \boldsymbol{\varphi}_e, \qquad\qquad \phi^\mathrm{h} = \boldsymbol{N}^\phi \cdot \boldsymbol{\phi}_e\,, \tag{27}$$

and introduce the shape functions in matrix form as

$$\boldsymbol{N}^{\varphi} = \begin{bmatrix} N_1^{\varphi} & 0 & 0 & \dots & N_{n_{\text{en}}^{\varphi}}^{\varphi} & 0 & 0 \\ 0 & N_1^{\varphi} & 0 & \dots & 0 & N_{n_{\text{en}}^{\varphi}}^{\varphi} & 0 \\ 0 & 0 & N_1^{\varphi} & \dots & 0 & 0 & N_{n_{\text{en}}^{\varphi}}^{\varphi} \end{bmatrix}, \quad \boldsymbol{N}^{\phi} = \begin{bmatrix} N_1^{\phi} & \dots & N_{n_{\text{en}}^{\phi}}^{\phi} \end{bmatrix}. \tag{28}$$

Similarly, the spatial gradients $\nabla_{\boldsymbol{x}}\boldsymbol{\varphi}$ and $\nabla_{\boldsymbol{x}}\phi$ of the placement $\boldsymbol{\varphi}$ and the non-local damage variable ϕ are approximated by means of

$$\nabla_{\boldsymbol{x}}\boldsymbol{\varphi}^{\text{h}} = \boldsymbol{B}^{\varphi} \cdot \boldsymbol{\varphi}_e, \qquad \nabla_{\boldsymbol{x}}\boldsymbol{\phi}^{\text{h}} = \boldsymbol{B}^{\phi} \cdot \boldsymbol{\phi}_e, \tag{29}$$

where we express the spatial gradients of the shape functions $N^{\alpha}_{n^{\alpha}_{\text{en}},x} := \nabla_x N^{\alpha}_{n^{\alpha}_{\text{en}}}$ in matrix form as

$$\boldsymbol{B}^{\varphi} = \begin{bmatrix} N_{1,x}^{\varphi} & 0 & 0 & \dots & N_{n_{\text{en}}^{\varphi},x}^{\varphi} & 0 & 0 \\ 0 & N_{1,y}^{\varphi} & 0 & \dots & 0 & N_{n_{\text{en}}^{\varphi},y}^{\varphi} & 0 \\ 0 & 0 & N_{1,z}^{\varphi} & \dots & 0 & 0 & N_{n_{\text{en}}^{\varphi},z}^{\varphi} \\ N_{1,y}^{\varphi} & N_{1,x}^{\varphi} & 0 & \dots & N_{n_{\text{en}}^{\varphi},y}^{\varphi} & N_{n_{\text{en}}^{\varphi},x}^{\varphi} & 0 \\ N_{1,z}^{\varphi} & 0 & N_{1,x}^{\varphi} & \dots & N_{n_{\text{en}}^{\varphi},z}^{\varphi} & 0 & N_{n_{\text{en}}^{\varphi},x}^{\varphi} \\ 0 & N_{1,z}^{\varphi} & N_{1,y}^{\varphi} & \dots & 0 & N_{n_{\text{en}}^{\varphi},z}^{\varphi} & N_{n_{\text{en}}^{\varphi},y}^{\varphi} \end{bmatrix}, \quad \boldsymbol{B}^{\phi} = \begin{bmatrix} N_{1,x}^{\phi} & \dots & N_{n_{\text{en}}^{\phi},x}^{\phi} \\ N_{1,y}^{\phi} & \dots & N_{n_{\text{en}}^{\phi},y}^{\phi} \\ N_{1,z}^{\phi} & \dots & N_{n_{\text{en}}^{\phi},z}^{\phi} \end{bmatrix}. \tag{30}$$

According to the Bubnov-Galerkin method, we apply an analogous approximation for the variations of field variables and corresponding gradients. The discretised weak form on element level is characterised by the difference of the element-specific internal and external virtual work related terms $\delta_{\boldsymbol{\varphi}} W_{e\,\text{int}}$ and $\delta_{\boldsymbol{\varphi}} W_{e\,\text{ext}}$, so that

$$\delta_{\boldsymbol{\varphi}} W_e = \delta_{\boldsymbol{\varphi}} W_{e\,\text{int}} - \delta_{\boldsymbol{\varphi}} W_{e\,\text{ext}} = \boldsymbol{0} \quad \forall\ \delta\boldsymbol{\varphi}_e, \tag{31}$$

$$\delta_{\boldsymbol{\phi}} W_e = \delta_{\boldsymbol{\phi}} W_{e\,\text{int}} - \delta_{\boldsymbol{\phi}} W_{e\,\text{ext}} = \boldsymbol{0} \quad \forall\ \delta\boldsymbol{\phi}_e, \tag{32}$$

where the discrete representations take the following forms

$$\delta_{\boldsymbol{\varphi}} W_{e\,\text{int}} = \delta\boldsymbol{\varphi}_e \cdot \boldsymbol{f}^{\varphi}_{e\,\text{int}} \quad \text{with} \quad \boldsymbol{f}^{\varphi}_{e\,\text{int}} = \int_{\mathcal{B}_t} \boldsymbol{B}^{\varphi\text{t}} \cdot \boldsymbol{\sigma}^{\text{v}}\, \mathrm{d}v\,, \tag{33}$$

$$\delta_{\boldsymbol{\varphi}} W_{e\,\text{ext}} = \delta\boldsymbol{\varphi}_e \cdot \boldsymbol{f}^{\varphi}_{e\,\text{ext}} \quad \text{with} \quad \boldsymbol{f}^{\varphi}_{e\,\text{ext}} = \int_{\mathcal{B}_t} \boldsymbol{N}^{\varphi\text{t}} \cdot \bar{\boldsymbol{b}}\, \mathrm{d}v + \int_{\partial\mathcal{B}_t} \boldsymbol{N}^{\varphi\text{t}} \cdot \bar{\boldsymbol{t}}\, \mathrm{d}a\,, \tag{34}$$

$$\delta_{\boldsymbol{\phi}} W_{e\,\text{int}} = \delta\boldsymbol{\phi}_e \cdot \boldsymbol{f}^{\phi}_{e\,\text{int}} \quad \text{with} \quad \boldsymbol{f}^{\phi}_{e\,\text{int}} = \int_{\mathcal{B}_t} \boldsymbol{B}^{\phi\text{t}} \cdot \boldsymbol{y}\, \mathrm{d}v\,, \tag{35}$$

$$\delta_{\boldsymbol{\phi}} W_{e\,\text{ext}} = \delta\boldsymbol{\phi}_e \cdot \boldsymbol{f}^{\phi}_{e\,\text{ext}} \quad \text{with} \quad \boldsymbol{f}^{\phi}_{e\,\text{ext}} = \int_{\mathcal{B}_t} \boldsymbol{N}^{\phi\text{t}}\, y\, \mathrm{d}v\,, \tag{36}$$

and Voigt-notation is denoted by the superscript v. Applying the fundamental lemma of calculus of variations, this results in the residual form of the present coupled problem

$$\boldsymbol{r}_e^{\varphi} = \boldsymbol{f}_{e\,\text{int}}^{\varphi} - \boldsymbol{f}_{e\,\text{ext}}^{\varphi} = \boldsymbol{0}\,, \tag{37}$$

$$\boldsymbol{r}_e^{\phi} = \boldsymbol{f}_{e\,\text{int}}^{\phi} - \boldsymbol{f}_{e\,\text{ext}}^{\phi} = \boldsymbol{0}\,. \tag{38}$$

4.2 Linearisation

The governing system of Eqs. (37) and (38) is coupled with respect to the motion $\boldsymbol{\varphi}$ and the non-local damage field ϕ, which itself is linked to the local damage variable κ by means of the penalty term. To solve this highly non-linear system of equations, we use an incremental-iterative Newton-Raphson scheme where we subsequently omit the subscript index $n+1$ associated with time t_{n+1} for the sake of readability. A Taylor series expansion around the solution at the current iteration step k up to the linear term gives

$$\boldsymbol{r}_{e\,k+1}^{\varphi} = \boldsymbol{r}_{e\,k}^{\varphi} + \frac{\mathrm{d}\boldsymbol{r}^{\varphi}}{\mathrm{d}\boldsymbol{\varphi}} \cdot \Delta\boldsymbol{\varphi} + \frac{\mathrm{d}\boldsymbol{r}^{\varphi}}{\mathrm{d}\boldsymbol{\phi}} \cdot \Delta\boldsymbol{\phi} = \boldsymbol{0}\,, \tag{39}$$

$$\boldsymbol{r}_{e\,k+1}^{\phi} = \boldsymbol{r}_{e\,k}^{\phi} + \frac{\mathrm{d}\boldsymbol{r}^{\phi}}{\mathrm{d}\boldsymbol{\varphi}} \cdot \Delta\boldsymbol{\varphi} + \frac{\mathrm{d}\boldsymbol{r}^{\phi}}{\mathrm{d}\boldsymbol{\phi}} \cdot \Delta\boldsymbol{\phi} = \boldsymbol{0}\,. \tag{40}$$

Herein, the increments $\Delta\boldsymbol{\varphi} = \boldsymbol{\varphi}_{k+1} - \boldsymbol{\varphi}_k$ and $\Delta\boldsymbol{\phi} = \boldsymbol{\phi}_{k+1} - \boldsymbol{\phi}_k$ represent the difference of the discrete nodal degrees of freedom at iteration-step $k+1$ and k. Assuming 'dead loads', we deduce the element-specific sub-matrices of the Jacobian as

$$\boldsymbol{K}_e^{\varphi\varphi} = \frac{\mathrm{d}\boldsymbol{r}^{\varphi}}{\mathrm{d}\boldsymbol{\varphi}} = \int_{\mathcal{B}_t^e} \left[\boldsymbol{B}^{\varphi\,\mathrm{t}} \cdot \boldsymbol{e}^{\mathrm{v}} \cdot \boldsymbol{B}^{\varphi} + [\boldsymbol{G}^{\varphi} \cdot \boldsymbol{\sigma} \cdot \boldsymbol{G}^{\varphi\,\mathrm{t}}] \odot \boldsymbol{I} \right] \mathrm{d}v\,, \tag{41}$$

$$\boldsymbol{K}_e^{\varphi\phi} = \frac{\mathrm{d}\boldsymbol{r}^{\varphi}}{\mathrm{d}\boldsymbol{\phi}} = \int_{\mathcal{B}_t^e} \boldsymbol{B}^{\varphi\,\mathrm{t}} \cdot \left[\frac{\mathrm{d}\boldsymbol{\sigma}}{\mathrm{d}\phi} \right]^{\mathrm{v}} \cdot \boldsymbol{N}^{\phi}\, \mathrm{d}v\,, \tag{42}$$

$$\boldsymbol{K}_e^{\phi\varphi} = \frac{\mathrm{d}\boldsymbol{r}^{\phi}}{\mathrm{d}\boldsymbol{\varphi}} = \int_{\mathcal{B}_t^e} \boldsymbol{N}^{\phi\,\mathrm{t}} \cdot \left[2\, \frac{\mathrm{d}y}{\mathrm{d}\boldsymbol{g}} \right]^{\mathrm{v}} \cdot \boldsymbol{B}^{\varphi}\, \mathrm{d}v\,, \tag{43}$$

$$\boldsymbol{K}_e^{\phi\phi} = \frac{\mathrm{d}\boldsymbol{r}^{\phi}}{\mathrm{d}\boldsymbol{\phi}} = \int_{\mathcal{B}_t^e} \left[\boldsymbol{N}^{\phi\,\mathrm{t}} \left[\frac{\mathrm{d}y}{\mathrm{d}\phi} \right]^{\mathrm{v}} \cdot \boldsymbol{N}^{\phi} + \boldsymbol{B}^{\phi\,\mathrm{t}} \cdot \left[\frac{\mathrm{d}\boldsymbol{y}}{\mathrm{d}\nabla_{\boldsymbol{x}}\phi} \right]^{\mathrm{v}} \cdot \boldsymbol{B}^{\phi} \right] \mathrm{d}v\,, \tag{44}$$

allowing us to express the geometrical contribution of $\boldsymbol{K}^{\varphi\varphi}$ conveniently by means of the Kronecker product $\odot$. Furthermore, $\boldsymbol{e}$, $\mathrm{d}\boldsymbol{\sigma}/\,\mathrm{d}\phi$, $2\,\mathrm{d}y/\,\mathrm{d}\boldsymbol{g}$, $\mathrm{d}y/\,\mathrm{d}\phi$ and $\mathrm{d}\boldsymbol{y}/\,\mathrm{d}\nabla_{\boldsymbol{x}}\phi$

denote partitions of the consistent tangent-moduli as specified in Waffenschmidt et al. [16]. This results in the following symmetric linearised system of equations

$$\begin{bmatrix} \boldsymbol{r}_e^{\varphi} \\ \boldsymbol{r}_e^{\phi} \end{bmatrix} + \begin{bmatrix} \boldsymbol{K}_e^{\varphi\varphi} & \boldsymbol{K}_e^{\varphi\phi} \\ \boldsymbol{K}_e^{\phi\varphi} & \boldsymbol{K}_e^{\phi\phi} \end{bmatrix} \cdot \begin{bmatrix} \Delta\boldsymbol{\varphi}_e \\ \Delta\boldsymbol{\phi}_e \end{bmatrix} = \begin{bmatrix} \boldsymbol{0} \\ \boldsymbol{0} \end{bmatrix} \tag{45}$$

for the determination of the unknown increments of the element degrees of freedom $\Delta\boldsymbol{\varphi}_e$ and $\Delta\boldsymbol{\phi}_e$. The assembly of all elements results in the global linearised system of equations in the kth iteration step

$$\boldsymbol{r} + \boldsymbol{K} \cdot \Delta\boldsymbol{d} = \boldsymbol{0} \quad \text{with} \quad \Delta\boldsymbol{d} = \boldsymbol{d}_{k+1} - \boldsymbol{d}_k, \tag{46}$$

with $\boldsymbol{K}$ being the global tangent stiffness matrix, $\Delta\boldsymbol{d}$ the global incremental vector of degrees of freedom, $\boldsymbol{r}$ the global residual vector including the internal and external system loads.

5 Residual Stresses

The incorporation of residual stresses is of key importance within the modelling and simulation of soft biological tissues. Different concepts have been discussed in the literature to account for these equilibrated stress contributions present in the absence of external loading. The procedure employed here to incorporate such effects is based on a multiplicative composition of the total deformation gradient as discussed by Johnson and Hoger [7] and Holzapfel et al. [6]. To be specific, we consider a stress-free reference configuration $\mathcal{B}_0$, a load-free residually stressed configuration $\mathcal{B}_{\text{res}}$ and a current configuration $\mathcal{B}_t$, cf. Fig. 1. In this regard, we—on the one hand—introduce a deformation gradient-type tensor $\boldsymbol{F}_{\text{res}}$ which transforms line elements from the stress-free reference configuration $\mathcal{B}_0$ to the residually stressed configuration $\mathcal{B}_{\text{res}}$ and—on the other hand—another deformation gradient-type tensor $\boldsymbol{F}_{\text{p}}$ which transforms line elements from the residually stressed configuration $\mathcal{B}_{\text{res}}$ to the current configuration $\mathcal{B}_t$. The resulting deformation gradient tensor can then be written in terms of cylindrical base vectors as

$$\boldsymbol{F} = \boldsymbol{F}_{\text{p}} \cdot \boldsymbol{F}_{\text{res}} = \lambda_i \, \boldsymbol{e}_i \otimes \boldsymbol{E}_i \,|_{i \in \{r,\theta,z\}} \ . \tag{47}$$

For a perfect cylindrical geometry, the radial and the circumferential principal stretches can be expressed in terms of spatial cylindrical coordinates $\{r, \theta, z\}$ as $\lambda_r = R/[r\,k\,\lambda_z]$ and $\lambda_\theta = r\,k/R$, where $R = \sqrt{k\,\lambda_z[r^2 - r_{\text{i}}^2] + R_{\text{i}}^2}$. In these relations, we prescribe the inner radius of the closed configuration r_{i}, the inner radius of the opened configuration R_{i}, the opening angle α, and the axial residual stretch λ_z, and introduce the opening angle parameter as $k = 2\pi/[2\pi - \alpha]$. Composition (47) can

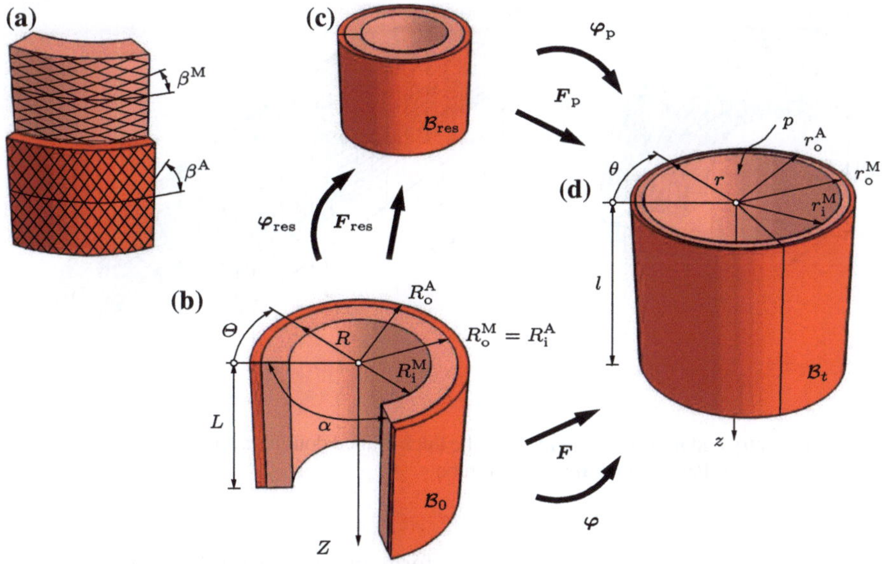

Fig. 1 Deformation modes of a double-layered thick-walled cylindrical tube, cf. Holzapfel et al. [6]: bending (opening angle α), inflation (internal pressure p), and extension (axial stretch λ_z); **a** segment of the arterial wall consisting of media (M) and adventitia (A) reinforced by two families of fibres with fibre angles β^{M}, β^{A} defined in **b** a stress-free reference configuration $\mathcal{B}_0$; **c** residually stressed but load-free configuration $\mathcal{B}_{\mathrm{res}}$; **d** residually stressed and loaded current configuration $\mathcal{B}_t$. (Adopted from Waffenschmidt and Menzel [15] with kind permission of Elsevier.)

conveniently be used to model residual stresses within finite element formulations for the simulation of tube-like boundary value problems; see Sect. 6.1.

6 Numerical Examples

In order to highlight the mechanical modelling capabilities as well as the computational performance of the presented model, we discuss some illustrative three-dimensional finite element examples. To be specific, we study the degradation of a fibre-reinforced artery-like tube subjected to internal pressure. As a crucial aspect of this contribution, residual stresses are incorporated by means of the technique described in Sect. 5.

6.1 Reproduction of the Opening Angle Experiment

Residual circumferential stresses can be revealed by the *opening angle experiment*, where a short ring of an artery is cut in radial direction. The residual stresses through

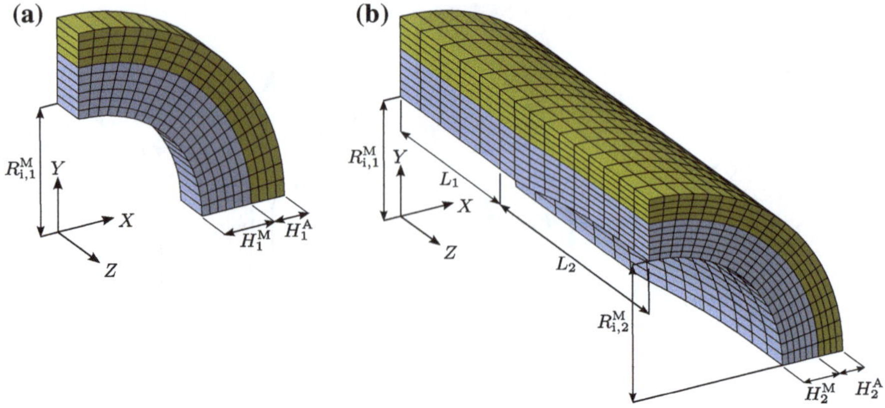

Fig. 2 Geometry and finite element mesh of the thick-walled double-layered tube, where only one quarter is shown. **a** Perfect tube. **b** Perturbed tube

the artery cause the ring to spring open to form an open sector. We now carry out a numerical reproduction of the opening angle experiment for a thick-walled double-layered tube, in order to assess whether the method provided in Sect. 5 yields accurate results. This technique allows for the inclusion of different opening angles for the different arterial layers within only one simulation. For this purpose, it is necessary to create an initial deformation gradient $\boldsymbol{F}_{\mathrm{res}}$ at each Gauss-point of a related finite element mesh. Note that $\boldsymbol{F}_{\mathrm{res}}$ must be inhomogeneously distributed as also the residual stresses are. This deformation gradient field has to be provided in terms of the underlying coordinate system used by the finite element code which is typically described by a Cartesian basis. The procedure to calculate the initial residual strain field by means of the deformation gradient tensor $\boldsymbol{F}_{\mathrm{res}}$ is described in detail in, for instance, Waffenschmidt [14].

The geometrical setting essentially reflects the geometry of a healthy coronary artery. We use two different geometries, i.e. (i) a perfect tube with constant thickness and (ii) a perturbed tube with variable thickness, see Fig. 2. The perturbed tube is represented by two different cross-sections. A first cross-sections (length L_1) with constant inner radius $R^{\mathrm{M}}_{\mathrm{i},1}$ and a second cross-sections with different inner radii $R^{\mathrm{M}}_{\mathrm{i},1}$ and $R^{\mathrm{M}}_{\mathrm{i},2}$. The outer radius R_{o} remains constant over the whole length. The functional relation of $R^{\mathrm{M}}_{\mathrm{i}}$ for $Z \in [L_1, L_2]$ is provided by

$$R^{\mathrm{M}}_{\mathrm{i}}(\bar{Z}) = -0.904\,\bar{Z}^5 + 3.913\,\bar{Z}^4 - 5.756\,\bar{Z}^3 + 3.259\,\bar{Z}^2 - 0.317\,\bar{Z} + 0.006\,, \quad (48)$$

where $\bar{Z} = [Z - L_1]/L_2$ such that $R^{\mathrm{M}}_{\mathrm{i}}(Z = L_1) = R^{\mathrm{M}}_{\mathrm{i},1}$ and $R^{\mathrm{M}}_{\mathrm{i}}(Z = L_1 + L_2) = R^{\mathrm{M}}_{\mathrm{i},2}$. The geometrical, structural and material parameters for both geometries are summarised in Table 2. We choose four different sets of the opening angles α^{n}, whereas we neglect the influence of the axial residual stretch at this stage, i.e. $\lambda_z = 1.0$. The initial orientations of the fibres are assumed as $\boldsymbol{a}^{\mathrm{n}}_{0\,1,2} = \sin\left(\beta^{\mathrm{n}}\right)\boldsymbol{e}_Z \pm$

Table 2 Set of parameters as used in the numerical examples (n = M for the media and n = A for the adventitia)

Type	Symbol	Value		Unit
		Media	Adventitia	
Geometrical	$R^{\mathrm{n}}_{\mathrm{i},1}$	1.35	1.89	mm
	H^{n}_{1}	0.54	0.36	mm
	$R^{\mathrm{n}}_{\mathrm{i},2}$	1.55	1.97	mm
	H^{n}_{2}	0.42	0.28	mm
	L_1	2.0		mm
	L_2	3.0		mm
Structural	β^{n}	21.700	62.260	deg
	$\varkappa^{\mathrm{n}}$	0.19	0.036	–
	α^{n}	{0.0, 45.0, 60.0, 120.0}	{0.0, 90.0, 120.0, 160.0}	deg
Elastic	$\mu^{\mathrm{n}}_{\mathrm{e}}$	27.0	2.7	kPa
	$\kappa^{\mathrm{n}}_{\mathrm{e}}$	2700.0	270.0	kPa
	k^{n}_{1}	0.64	5.1	kPa
	k^{n}_{2}	7.08	30.8	–
Regularisation	c_{d}	{0.08, 0.5, 3.0}		$\mathrm{kPa}^{-1}\mathrm{mm}^2$
	β_{d}	1000.0		kPa^{-1}
Damage	η_{d}	0.25		kPa^{-1}
	κ_{d}	5.0		kPa

$\cos\left(\beta^{\mathrm{n}}\right) \boldsymbol{e}_{\Theta}$ with n = M for the media and n = A for the adventitia. The perfect tube is discretised by 800, the perturbed tube by 12,000 finite elements. Since we prescribe the radii $R^{\mathrm{n}}_{\mathrm{i}}$ of the opened configuration and the opening angles α^{n} as input parameters to create the residual strain field, we expect to reproduce these values at least approximately as well as a complete stress release at the end of the simulation. Basically the FE-simulation of the opening angle experiment is subdivided into a sequence of two calculation steps:

1. *Application of the residual strain field* at Gauss point level: apply homogeneous Dirichlet boundary conditions at all nodes in circumferential and axial direction.
2. *Radial cut*: apply homogeneous Dirichlet boundary conditions at two nodes in radial direction, at all nodes located at the cut in circumferential direction and at all nodes at back face in axial direction.

In this regard, all boundary conditions are removed from one step to another and re-specified for those boundary conditions that are to be retained. If a boundary condition is removed, it will be replaced by a concentrated force equal to the reaction force calculated at the restrained degree of freedom at the end of the previous step. This concentrated force will then be linearly reduced to zero over the period of the step. In this case, both layers are separated from each other in order to give an illustrative view

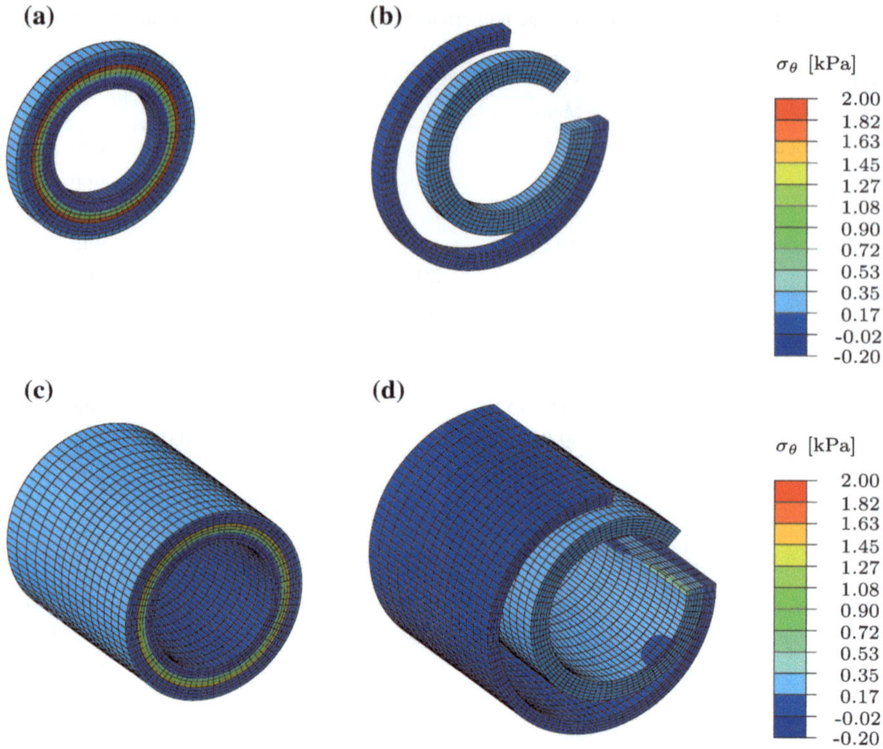

Fig. 3 Reproduction of the opening angle experiment with prescribed opening angles $\alpha^{\text{M}} = 45.0\,[\text{deg}]$, $\alpha^{\text{A}} = 90.0\,[\text{deg}]$. Circumferential Cauchy stress σ_θ after **a** the application of the residual strain field, **b** the radial cut

of the effect of different prescribed opening angles and radii which, for the purpose of illustration, are set to values of $R_{\text{i}}^{\text{M}} = 1.5\,[\text{mm}]$, $R_{\text{i}}^{\text{A}} = 2.5\,[\text{mm}]$, $\alpha^{\text{M}} = 45.0\,[\text{deg}]$ and $\alpha^{\text{A}} = 90.0\,[\text{deg}]$. As a consequence of the initial imposition of the residual strain field, we obtain a stretching and warping of the cross-section of the tube. As both layers are separated from each other, this consequently results in an undesired penetration. In order to avoid this, a contact condition is prescribed between outer media and inner adventitia in the first simulation step.

Figure 3 illustrates the contours of the circumferential Cauchy stress σ_θ at different configurations $\mathcal{B}_0$ and $\mathcal{B}_{\text{res}}$ of the simulation for the perfect tube (Fig. 3a, b) and the perturbed tube (Fig. 3c, d). It becomes apparent that the technique described above is able to appropriately reproduce the prescribed input parameters, so that $R_{\text{i sim}}^{\text{M}} \approx R_{\text{i}}^{\text{M}}$, $R_{\text{i sim}}^{\text{A}} \approx R_{\text{i}}^{\text{A}}$, $\alpha_{\text{sim}}^{\text{M}} \approx \alpha^{\text{M}}$ and $\alpha_{\text{sim}}^{\text{A}} \approx \alpha^{\text{A}}$ and $\sigma_\theta \approx 0$. Furthermore, the stress levels approximately agree for both cases. This underlines that the assumption of a perfect cylindrical tube is—in this case—a valid approximation for residual stress imposition of a slightly perturbed tube as used in this contribution.

6.2 Inflation of the Perturbed Tube

We continue with a study of the pressure-driven inflation problem of the perturbed tube. First, we focus on the elastic response and the effect of the residual stresses on the overall behaviour of the structure. Due to the symmetry of the inflation problem, it is sufficient to consider only one quarter of the perturbed tube resulting in 3,000 finite elements, as shown in Fig. 2.

6.2.1 Elastic Response

Figure 4a shows the load-displacement diagram in terms of the internal pressure p over the radial displacement $u_{r\text{i}}$ at the inner radius of the tube at $z = L_1 + L_2$. Two different elastic solutions are shown: the isotropic neo-Hookean matrix in the lower dotted curves, and the overall elastic response composed of the neo-Hookean matrix *and* the fibre reinforced material in the upper solid curves. These curves give a good illustration of the highly non-linear material behaviour especially at large strains. As a consequence of the residual stresses, we basically obtain two significant effects. First, we observe a shift towards negative radial displacements. Secondly, we observe a shift towards higher pressure values. Both effects are directly related to the contraction of the tube due to the residual stresses. The contraction basically implies a change in the configuration in which the pressure is applied, i.e. for a reduced inner radius a larger pressure must be applied in order to obtain the same radial displacement. This is qualitatively in line with the solution for thick-walled cylinders under internal pressure in linear elasticity.

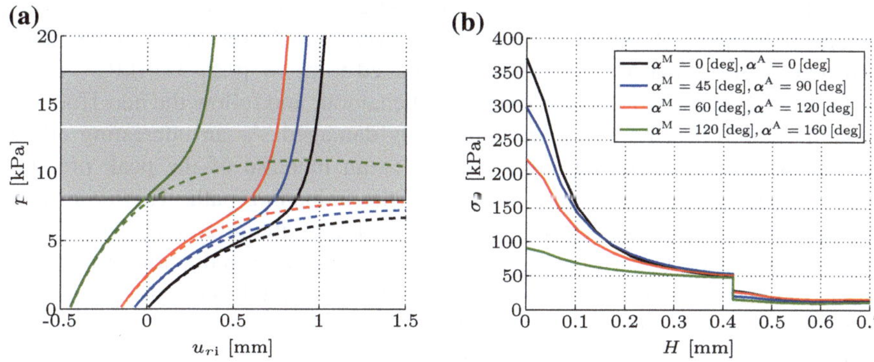

Fig. 4 Perturbed tube subjected to internal pressure and residual stresses. Variation of different sets of opening angles α^{n}. *Solid lines* represent the response of the fibre-reinforced material, *dotted lines* represent the response of the neo-Hookean matrix at $z = L_1 + L_2$. **a** Pressure p versus inner radial displacement $u_{r\text{i}}$, where the *grey region* represents the physiological pressure range; **b** distribution of the circumferential Cauchy stress σ_θ along the undeformed thickness H at $p = 13.33$ [kPa]

The residual stresses also affect the distribution of the circumferential stresses over the undeformed tube thickness H, see Fig. 4b for a physiological (mean) internal pressure of $p = 13.33$ [kPa]. Apart from the discontinuity in the stress distribution which is due to the different material properties in media and adventitia, we observe that the stresses are significantly homogenised over the tube thickness and that their maximum value is dramatically reduced. This effect has a direct impact on the characteristic pressure range in which inelastic damage processes start to evolve as a result of high strain energy and stress levels. This effect is studied in more detail in the subsequent section.

6.2.2 Inelastic Response

Next, we study the degradation of the fibre-reinforced perturbed tube. As we are dealing with a force-driven boundary value problem, we use an arc-length method in order to trace the equilibrium path in case of snap-back-related instabilities. A maximum inner radial displacement of $u_{r\mathrm{i}} = 1.5$ [mm] is set as a limiting point in the load-displacement diagram shown in Fig. 5. Furthermore, we choose an appropriate set of material parameters to initiate the damage process as soon as the pressure exceeds the systolic blood pressure of approximately $p \approx 20.0$ [kPa]. The results are indicated in Fig. 5, where we adjust the pressure plot-range to a maximum value of 70.0 [kPa]. As a consequence, the stiffening due to the residual-stress-induced contraction is no longer visible but still present. The lower solid black curve represents the elastic response of the neo-hookean ground substance, the upper solid black curve represents the elastic response of the fibre-reinforcement material. The colored lines represent the response of the damaging structure. As only the fibres are assumed to be affected by the damage, the associated response is always located between these two black lines representing the elastic response. Upon successive damage of the fibres, the structural response converges to the response of the undamaged neo-Hookean matrix. The curves associated to $c_\mathrm{d} = \{0.5,\ 3.0\}[\mathrm{kPa}^{-1}\ \mathrm{mm}^2]$ in Fig. 5a–c show a characteristic snap-back behaviour and follow the neo-Hookean unloading path once the fibres are completely damaged. As an interesting effect, the incorporation of residual stresses leads to an increase of the peak pressure before the overall structural response enters the unloading path. This is a direct consequence of the homogenisation-tendency of the circumferential stresses: as observed in Fig. 4b, larger opening angles α^n entail a larger reduction in the maximum circumferential stress σ_θ. Therefore, higher pressure levels can be sustained before the strain energy reaches the threshold to initiate the damage evolution. Particularly Fig. 5d shows that the structure exhibits almost no degradation for $c_\mathrm{d} = \{0.5,\ 3.0\}$ [$\mathrm{kPa}^{-1}\ \mathrm{mm}^2$].

Furthermore, we investigate the effect of the regularisation parameter c_d. Generally, smoother distributions for the non-local damage variable ϕ and the damage function f_d are obtained for larger values of c_d. This is illustrated in Fig. 6, where the damage function f_d is shown for $\alpha^\mathrm{M} = 45.0$ [deg], $\alpha^\mathrm{A} = 90.0$ [deg] and different values of c_d at a post-peak pressure of $p = 40.0$ [kPa], i.e. a point within the unloading

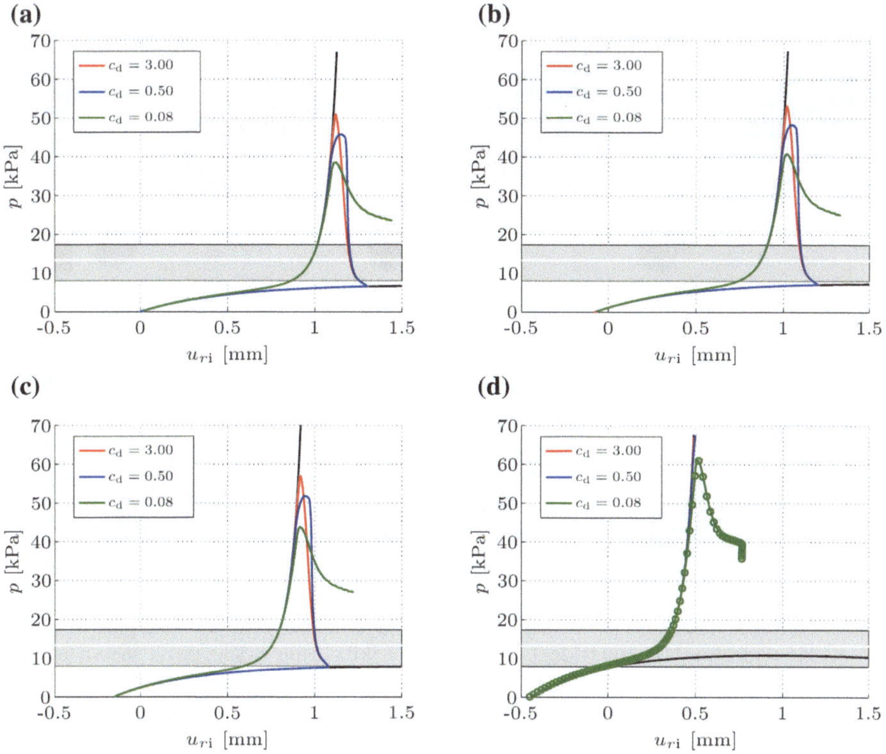

Fig. 5 Perturbed tube subjected to internal pressure and residual stresses. Variation of the regularisation parameter c_d in [kPa^{-1} mm^2] for different sets of opening angles α^n. The *upper* and *lower solid black lines* represent the purely elastic response of the fibre-reinforced material and the neo-Hookean matrix. The *colored lines* represent the damage response. **a** $\alpha^M = 0.0$, $\alpha^A = 0.0$, **b** $\alpha^M = 45.0$, $\alpha^A = 90.0$, **c** $\alpha^M = 60.0$, $\alpha^A = 120.0$, **d** $\alpha^M = 120.0$, $\alpha^A = 160.0$ where the *circular markers* are associated with a mesh of 12,000 elements. The *grey region* represents the physiological pressure range. Units of α^n in [deg]

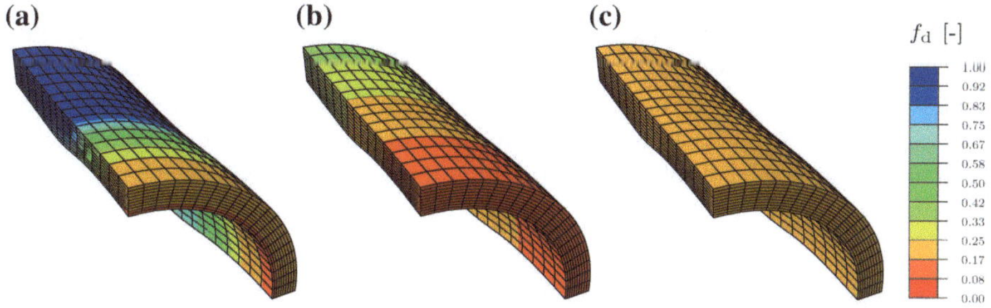

Fig. 6 Perturbed tube subjected to internal pressure and residual stresses. Contours of the damage function f_d for $\alpha^M = 45.0$ [deg], $\alpha^A = 90.0$ [deg] at a post-peak pressure of $p = 40.0$ [kPa]. Variation of f_d with **a** $c_d = 0.08$, **b** $c_d = 0.5$, **c** $c_d = 3.0$. Units of c_d in [kPa^{-1} mm^2]

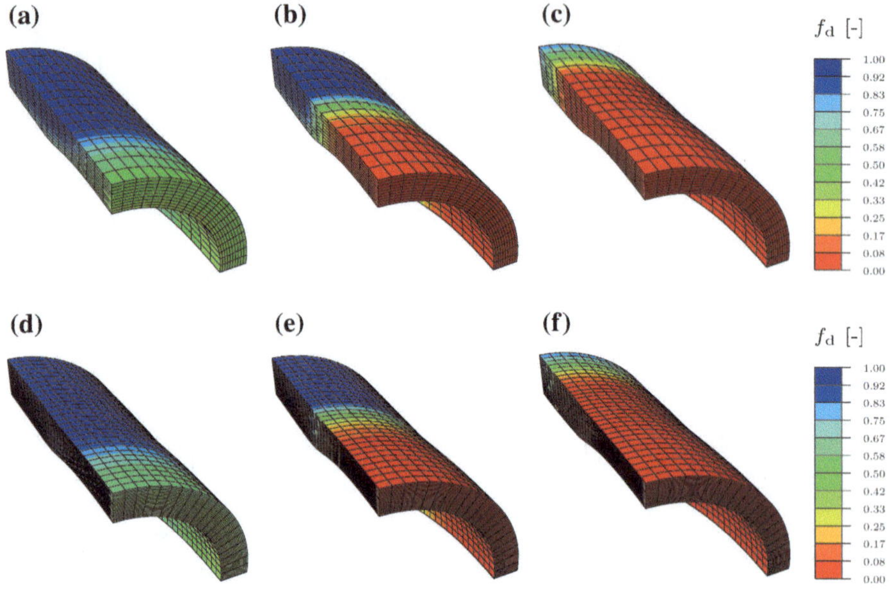

Fig. 7 Perturbed tube subjected to internal pressure and residual stresses. Contour plots of f_d with $\alpha^M = 120.0$ [deg], $\alpha^A = 160.0$ [deg], and $c_d = 0.08$ [kPa^{-1}mm^2]. Successive snapshots along the unloading path at pressure levels $p = \{61.08, 42.89, 36.33\}$ [kPa]. (**a**–**c**) associated with 3,000 elements; (**d**–**f**) associated with 12,000 elements

path. It becomes apparent that smaller values of c_d cause the damage distribution to be more localised towards the thinner portion of the tube.

Finally, Fig. 7 shows the evolution of the damage function f_d for the prestressed tube for $\alpha^M = 120.0$ [deg], $\alpha^A = 160.0$ [deg] and $c_d = 0.08$ [kPa^{-1}mm^2]. Three successive snapshots along the unloading path are shown in Fig. 7a–d which correspond to pressure levels of $p = \{61.08, 42.89, 36.33\}$ [kPa] for a discretisation of 3,000 elements. We observe that damage is first initiated at the thinnest section of the tube and later on evolves across the structure. Figure 7d–f show the same quantities for a finer discretisation of 12,000 elements. No substantial differences between both discretisations can be observed. Moreover Fig. 5d shows the load-displacement curves for both discretisations and again both curves almost completely coincide. This underlines the mesh-objective properties of the present gradient-enhanced damage model.

7 Summary

A well-known problem related to *local* continuum damage formulations is the possible ill-posedness of the underlying boundary value problem. As a consequence, finite element simulation results turn out to significantly depend on the discretisation. Typically, this is indicated by a vanishing localised damage zone, which is constrained only by the mesh resolution.

One possibility to regularise the problem, is the use of non-local continuum formulations. In this contribution, we present a non-local gradient-enhanced damage model within a finite strain setting. We implicitly regularise the problem by enhancing the local free energy with the gradient of the non-local damage variable and ensure the equivalence between the local and non-local damage field variable by a penalisation term incorporated within the free energy. The local elastic constitutive response is represented by a hyperelastic format, additively composed of an isotropic and an anisotropic part. The inelastic response is governed by a scalar $[1-d]$-damage approach, affecting the anisotropic elastic part only. As another major aspect of this contribution, we incorporate residual stresses which are crucial for the modelling of soft biological tissues such as arteries. The procedure employed here to incorporate these effects is based on a multiplicative composition of the total deformation gradient.

We present a biomechanics-related three-dimensional numerical example, i.e. the simulation of a thick-walled two-layered artery-like tube subjected to internal pressure and residual stresses. The highly non-linear elastic behaviour of the material is well reflected by the mechanical response of the tube within the physiological pressure range. For higher pressure values the material softens significantly, deviates from the elastic path and finally drops towards the elastic response of the undamaged neo-Hookean ground substance. Physically, this can be interpreted as a continuous degradation of the fibres contained in an undamaged matrix material. The damage evolution beyond the physiological range results in an increase of the tube's diameter at the same mean pressure. Consequently, at the expense of a loss in stiffness, these damage effects can be beneficially used in medical surgery to expand an artery's diameter and thereby restore the blood flow in atherosclerotic blood vessels for example. As an interesting consequence of the residual stresses and the related homogenised stress distribution over the tube thickness, we observe an increase of the peak pressure before the overall structural response enters the unloading path. In other words, due to the presence of residual stresses, arteries are able to sustain higher supraphysiological loads. This illustratively shows the beneficial impact of residual stresses on the overall behaviour of arteries.

It is important to mention that the assumption of an intact matrix and two equally damaging fibre families provides only a limited view of the real (bio)mechanical effects at this stage. Therefore, ongoing research work aims at extending the formulation to three global damage variables, including matrix-damage and individual damage for each fibre family.

References

1. Aifantis, E.: On the role of gradients in the localization of deformation and fracture. Int. J. Eng. Sci. **30**, 1279–1299 (1992)
2. Dimitrijević, B.J., Hackl, K.: A method for gradient enhancement of continuum damage models. Technische Mechanik **28**, 43–52 (2008)
3. Eringen, A. (ed.): Continuum Physics. Volume IV—Polar and Nonlocal Field Theories. Academic Press, New York (1976)
4. Eringen, A.: Nonlocal Continuum Field Theories. Springer, New York (2002)
5. Gasser, T., Ogden, R., Holzapfel, G.: Hyperelastic modelling of arterial layers with distributed collagen fibre orientations. J. R. Soc. Interface **3**, 15–35 (2006)
6. Holzapfel, G., Gasser, T., Ogden, R.: A new constitutive framework for arterial wall mechanics and a comparative study of material models. J. Elast. **61**, 1–48 (2000)
7. Johnson, B., Hoger, A.: The use of virtual configuration in formulating constitutive equations for residually stressed elastic materials. J. Elast. **41**, 177–215 (1995)
8. Kachanov, L.: Time of the rupture process under creep conditions. Izvestija Akademii Nauk Sojuza Sovetskich Socialisticeskich Respubliki (SSSR) Otdelenie Techniceskich Nauk (Moskra) vol. 8, pp. 26–31 (1958)
9. Krajcinovic, D., Lemaitre, J.: Continuum Damage Mechanics. CISM Courses and Lectures, vol. 295. Springer, New York (1987)
10. Lemaitre, J.: A Course on Damage Mechanics, 2nd edn. Springer, Berlin (1996)
11. Liebe, T., Steinmann, P., Benallal, A.: Theoretical and computational aspects of a thermodynamically constistent framework for geometrically linear gradient damage. Comput. Methods Appl. Mech. Eng. **190**, 6555–6576 (2001)
12. Menzel, A., Harrysson, M., Ristinmaa, M.: Towards an orientation-distribution-based multiscale approach for remodelling biological tissues. Comput. Methods Biomech. Biomed. Eng. **11**, 505–524 (2008)
13. Rogula, D. (ed.): Nonlocal Theory of Material Media. CISM Courses and Lectures, vol. 268. Springer, New York (1982)
14. Waffenschmidt, T.: Modelling and simulation of adaptation and degradation in anisotropic biological tissues. Ph.D. Thesis. Institut für Mechanik, Technische Universität Dortmund (2014). http://hdl.handle.net/2003/31797
15. Waffenschmidt, T., Menzel, A.: Extremal states of energy of a double-layered thick-walled tube—application to residually stressed arteries. J. Mech. Behav. Biomed. Mater. **29**, 635–654 (2014)
16. Waffenschmidt, T., Polindara, C., Menzel, A., Blanco, S.: A gradient-enhanced large-deformation continuum damage model for fibre-reinforced materials. Comput. Methods Appl. Mech. Eng. **268**, 801–842 (2014)

A Mechanically Stimulated Fracture Healing Model Using a Finite Element Framework

Alexander Sapotnick and Udo Nackenhorst

Abstract In this work a biochemical fracture healing model coupled with mechanical stimulation of stem cell differentiation is investigated. A finite element scheme is applied to the underlaying advection-diffusion-reaction problem, using the Time Discontinuous Galerkin and Finite Calculus method to ensure stability of the calculation. Strains within the callus region are computed and used for a characterization of the local mechanical demand and the resulting stimulation of the healing process. A theoretical axisymmetric model of a sheep osteotomy is implemented and results of the presented FEM approach are discussed. The repair progress will be determined by the interfragmentary movement (IFM) and the mean tissue densities.

Keywords Fracture healing · Bone repair · Biomechanics · Finite element

1 Introduction

The process of fracture healing, if successful, is a well-orchestrated biochemical response to the injury of the bone. Cells, mesenchymal stem cells in particular, are recruited from the surrounding tissue and are stimulated to produce new tissues in the callus region. The effect is a subsequent stabilisation and vascularization of the fracture site, until the normal load bearing capacity of the bone is regained. Growth factors play a major role in controlling the cell activities. They serve as attractants, drawing cells to locations of high growth factor concentration and stimulate stem cell differentiation. Additionally, the mechanical demand on the callus area influences the cellular functions. Exposure to high loads, premature or inappropriate loading inhibits bone growth and re-vascularization.

Typically, the process from the occurrence of the fracture to the bridging of the fracture gap and re-establishing of the load bearing capacity of the bone takes

A. Sapotnick (✉) · U. Nackenhorst
Institute of Mechanics and Numerical Mechanics, Leibniz Universität Hannover,
Appelstr. 9a, 30167 Hannover, Germany
e-mail: alexander.sapotnick@ibnm.uni-hannover.de

T. Lenarz and P. Wriggers (eds.), *Biomedical Technology*, Lecture Notes
in Applied and Computational Mechanics 74, DOI 10.1007/978-3-319-10981-7_3

about six to eight weeks. Six to eight months of healing is called delayed union and non-union if healing is delayed even further.

Mathematical models of the fracture healing process have been developed in the recent decades. The mechanical excitation and stimulation of the callus region has been the main focus of the majority of these models [1–5]. But also the biochemical influences have been highlighted in some of the simulations, like the re-establishing of blood vessels and nutrient supply [6]. In [7] a sophisticated model of biochemical fracture healing was presented, which was further refined in [8]. A mechanical stimulation was added to both models in [9, 10], respectively.

Here, a finite element approach for the model presented in [8] is proposed and coupled with a mechanical stimulus based on the strains in the callus area. This approach is investigated on an osteotomy model of sheep.

2 Mathematical Fracture Healing Model

The investigated model of biochemical fracture healing was developed by Bailón-Plaza and van der Meulen [7] and complemented in [8] by Geris et al., where additional equations for fibroblasts, connective tissue and angiogenesis were added. It will be reviewed here in abbreviated form, the interested reader is referred to the articles cited above and the authors own work [11] for a more detailed representation. The biochemical model describes the fracture healing process by incorporating five cell concentrations. The local change of concentration with time is given, in general, by an advection-diffusion-reaction equation

$$\frac{\partial c_i}{\partial t} = \nabla(D_i(m)\,\nabla c_i) - \nabla(C_i(g_k, m_j)\,c_i) + R_i(m_j, g_k), \tag{1}$$

where D_i is a diffusion parameter dependent on the total extracellular matrix density (ECM) ($m = \sum m_j$), C_i are advection coefficients representing chemo- and haptotactical movement controlled by growth factor concentrations and ECM densities. The index $i = m, f, c, b, v$ indicates the different cell types, stem cells, fibroblasts, chondrocytes, osteoblasts and endothelial cells, respectively. Furthermore, proliferation, differentiation and apoptosis are realized by reactive terms R_i, which are regulated by the biochemical milieu, i.e. matrix densities and growth factor concentrations.

The tissue development inside the callus is given by four extracellular matrix densities m_j, where $j = f, c, b, v$ denotes fibrous, cartilage, immature bone and vascular tissue, respectively. The production and, in case of connective tissue and cartilage the resorption, is represented by reaction equations

$$\frac{\partial m_j}{\partial t} = R_{prod,j}(1-m)c_j - R_{deg,j}(m_j)c_i. \tag{2}$$

Here again m denotes the total tissue density. Unlike the original model, where each tissue growth was capped by its individual local density through the term $(1 - m_j)$, a competition between the growing tissues was implemented by using the m in this case. Growth factors taking part in bone repair are grouped into osteogenic, chondrogenic and angiogenic growth factor concentrations. These concentrations, indicated by index $k = c, b, v$, are given by diffusion-reaction equations of the form

$$\frac{\partial g_k}{\partial t} = \nabla(D_k \nabla g_k) + R_{prod,k}(g_k, m_j)c_i - R_{deg,k}\, g_k, \tag{3}$$

where D_k is the diffusion coefficient and R_k are production and degradation terms. Growth factor production depends on the biochemical milieu and the respective cell populations c_i.

2.1 Mechanical Stimulation

An isotropic hyperelastic material law is used for the callus region, assuming a nearly incompressible material for the soft tissues. As a measure for local mechanical demand η within the callus the equivalent deviatoric strain ε'_V is used

$$\eta := \varepsilon'_V = \frac{2}{\sqrt{3}}\sqrt{\frac{1}{6}((\varepsilon'_1 - \varepsilon'_2)^2 + (\varepsilon'_2 - \varepsilon'_3)^2 + (\varepsilon'_3 - \varepsilon'_1)^2)}. \tag{4}$$

The level of mechanical excitation η then defines stimuli parameters ψ_l, with $l = b, c, f$ denoting osteogenic, chondrogenic and fibrogenic stimulation, as indicated in Fig. 1. The parameters ψ take on values between 0 for no and 1 for a maximal stimulation.

The implementation of the stimulations ψ_l is presented for the stem cell equation,

$$\begin{aligned}\frac{\partial c_m}{\partial t} &= \nabla[D_m \nabla c_m - C_m(\nabla g_b, \nabla g_v, \nabla m)c_m] + A_m c_m(1 - c_m) \\ &\quad - (\psi_b F_1 + \psi_c F_2 + \psi_f F_4)c_m.\end{aligned} \tag{5}$$

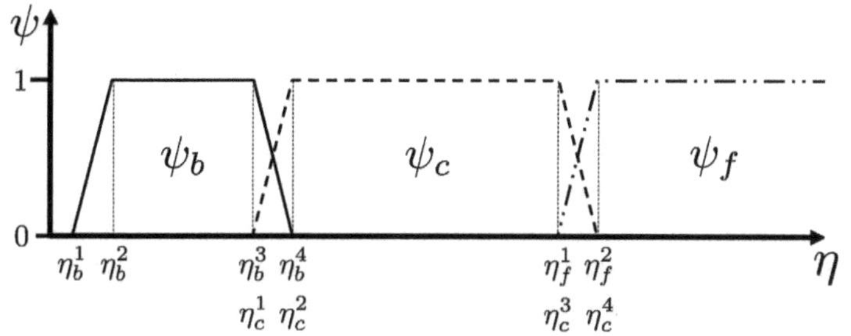

Fig. 1 Definition of stimulation parameter ψ by mechanical excitation levels η

Here D_m, C_m are diffusive and advective coefficients as above, A_m denotes the proliferation parameter and F_1, F_2 and F_4 are the differentiation coefficients for osteogenic, chondrogenic and fibrogenic differentiation. By multiplication of F_l with the respective ψ_l the intensity of the differentiation is scaled.

3 Methods

To solve the presented mathematical model the Finite Element method is employed. However, when a standard Galerkin approach is chosen, spurious oscillations often pollute the solutions of advection-diffusion-reaction equations. These oscillations occur in the vicinity of steep gradients in the solution, which are present from the beginning of the calculation (e.g. at Dirichlet boundaries) or build up during the calculation. Dominance of the advection or reaction terms often lead to steep gradients in the solution. This problem is well documented in literature and various techniques were developed to prevent oscillations.

In this work the Time Discontinuous Galerkin method (TDG) was combined with the finite calculus scheme (FIC) developed by Oñate [13, 14] to achieve a stable solution of the system of instationary advection-diffusion-reaction equations described above, for details it is referred to [11].

3.1 Simulation

As a geometrical model for the simulation a simplified axisymmetric representation of an osteotomy is chosen, as shown in Fig. 2. The dimensions of the callus region are chosen according to [4] to indicate a sheep metatarsal model. The fracture gap is set to 3 mm throughout the presented simulation.

The boundary conditions for biochemical simulation are shown on the right side of Fig. 2. They describe cell and growth factor reserviors in the surrounding tissue. Stem cells and fibroblasts are therefore recruited from the marrow cavity and soft tissue outside of the periosteal callus. Endothelial cells originate from the periosteum, endosteum and marrow. The source for chondrogenic growth factors is the fractured bone end and osteogenic growth factors invade the callus region from the periosteum and endosteum. The initial condition of the callus is chosen to be a connective tissue of low density, indicating the granulation tissue formed during the inflammation phase of bone healing. And a high concentration of angiogenic growth factor is present throughout the callus in accordance with observations documented in literature [15].

The mechanical simulation is executed at the start of each day of healing. At the beginning of the healing process a maximal displacement acts on the cortical bone. The displacement driven simulation continues until the reaction forces calculated on the upper boundary are equal to the maximal load of a sheep leg (i.e. about 600 N taken from [12]). Then this maximal force is applied to the former displacement

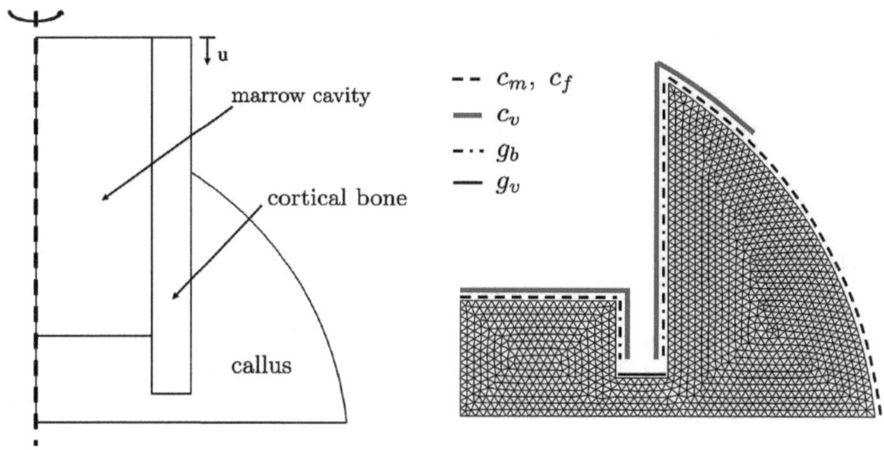

Fig. 2 Axisymmetric representation of sheep metatarsal fracture site (*left*), meshed callus region with indication of boundary conditions for cell and growth factor reserviors in the surrounding tissue (*right*)

Table 1 Mechanical properties of biological tissues

Tissue type	Young's modulus [MPa]	Poisson ratio
Cortical bone	1,7000[a]	0.3[a]
Immature bone	1,000[b]	0.3[b]
Cartilage	10[b]	0.47[b]
Connective tissue	2[b]	0.47[b]
Granulation tissue	0.18	0.47
Marrow	2[a]	0.3

a Prendergast et al. [2]
b Bailòn-Plaza et al. [9]

boundary as an axial load and with the stiffening of the callus the interfragmentary movement decreases.

The mechanical stimulus is calculated from the results of the mechanical simulation according to Sect. 2.1 and is applied to the biochemical simulation for the respective day of healing. The local Young's modulus is determined by a rule of mixture from the Young's moduli of the individual tissues shown in Table 1. The Poisson ratio of granulation tissue is set to near incompressibility, taking into account a high content of fluid. The excitation levels defining regions of equivalent strain, where stimulation of the various cell activities takes place are shown in Table 2. Additionally to stimulation of stem cell differentiation to osteoblasts, chondrocytes and fibroblasts, there is also a maximal strain assumed for the development of new blood vessels, given by the values for η_v^3 and η_v^4.

Table 2 Definition of excitation levels according to Fig. 1

η_b^1	η_b^2	η_b^3	η_b^4	η_c^1	η_c^2	η_c^3	η_c^4
0.0005	0.001	0.03	0.04	0.01	0.015	0.10	0.12
η_f^1	η_f^2	η_f^3	η_f^4			η_v^3	η_v^4
0.10	0.12	0.30	0.31			0.06	0.08

4 Results

The simulation was carried out, in a first step, for an initial displacement of 0.6 mm (indicating a total of 1.2 mm interfragmentary movement at the beginning of the healing period). A healing duration of 30 days was simulated. Figure 3 shows the

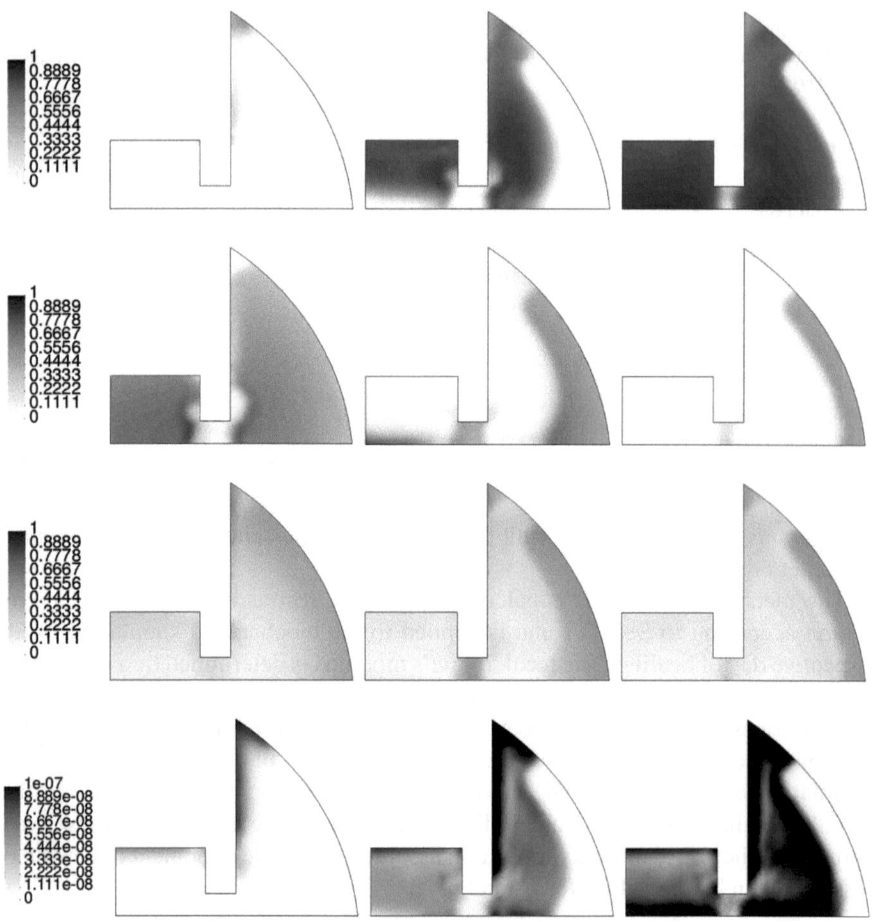

Fig. 3 Distribution of tissue density in callus region, *top to bottom*: immature bone, cartilage, fibrous tissue and vasculature, *left to right*: 10, 20 and 30 days of healing

distribution of fibrous, cartilage, immature bone and vascular tissue in the callus at certain times during the healing period. After ten days newly formed woven bone can be seen adjacent to the periosteum, while the remainder of the callus area is characterized by moderate to high densities of cartilage and connective tissues. The reconstitution of blood vessels can be observed in the endosteal callus region and in the vicinity of the periosteum. At the twentieth day the bridging of the fracture gap is already nearly completed. Soft tissues remain only in the interfragmental area

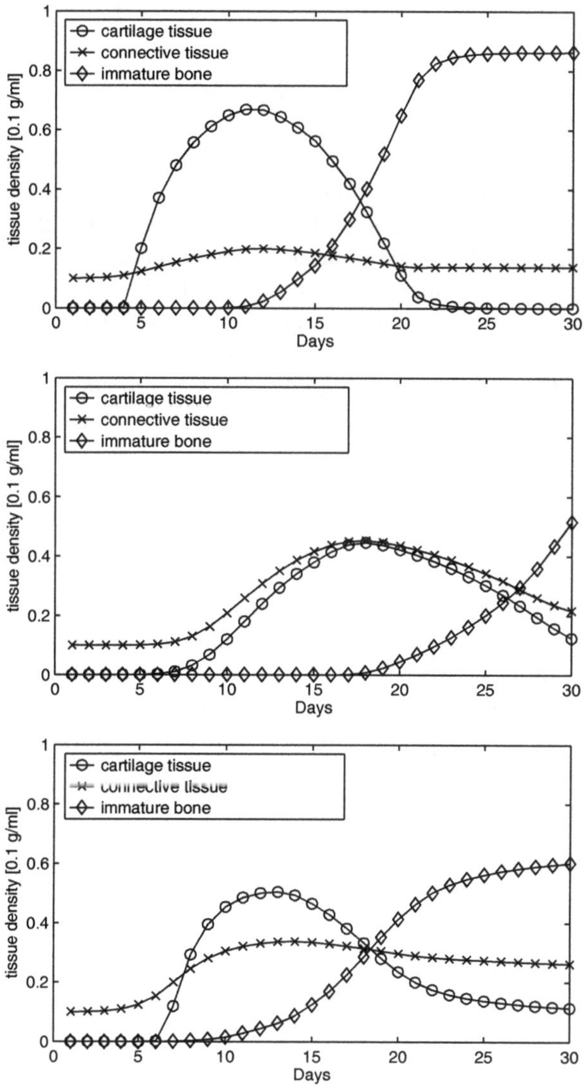

Fig. 4 Development of tissue densities; *top to bottom*: endosteal callus, interfragmental and periosteal callus area

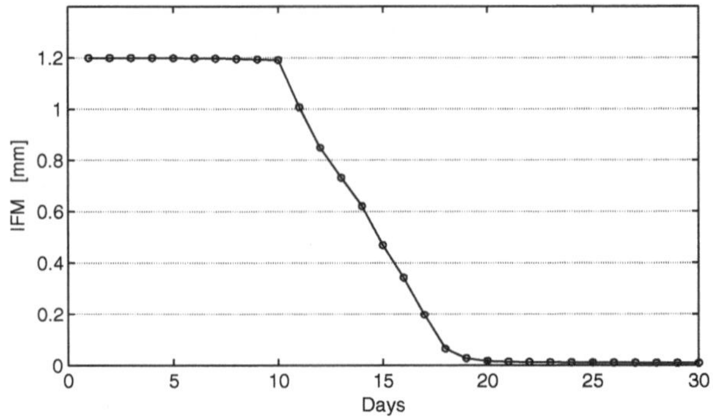

Fig. 5 Interfragmental Movement (IFM) during 30 days of healing

and at the outer margin of the periosteal callus. The blood supply has also been re-established. At the end of the simulation bony ingrowth into the fracture gap has taken place and the soft tissues have been replaced in the main part of the callus. Figure 4 shows the evolution of the mean tissue densities in the endosteal, interfragmental and periosteal callus. Formation of soft tissues in the fracture gap starts at around day ten, while production of immature bone can be seen to start not until day 17. The interfragmental movement determined at the end of the cortial bone in the fracture gap is shown in Fig. 5. It decreases rapidly after day ten and is reduced to nearly 0 mm at the end of the simulation.

4.1 Parameter Study

Calculations of the model with a variation of several parameters of the advection-diffusion-reaction equations were carried out to determine its sensitivity. Individual parameters were varied between 50 and 400 % of their original value. The focus here will be the variation of the reaction terms controlling the differentiation of stem cells.

A change of the differentiation to fibroblasts or osteoblasts has remarkedly little influence on the tissue development as shown in Figs. 6 and 7. The lack of differentiated cells is here balanced out by the logistic growth of the population, while an abundance of differentiated cells is capped when the maximal capacity of the population is reached.

A significant effect can be seen for the variation of the chondrocyte differentiation of the stem cells in Fig. 8. As can be expected an increase of the parameter leads to an earlier and more pronounced cartilage production, which in turn reduces the production of connective tissues and facilitates the ossification.

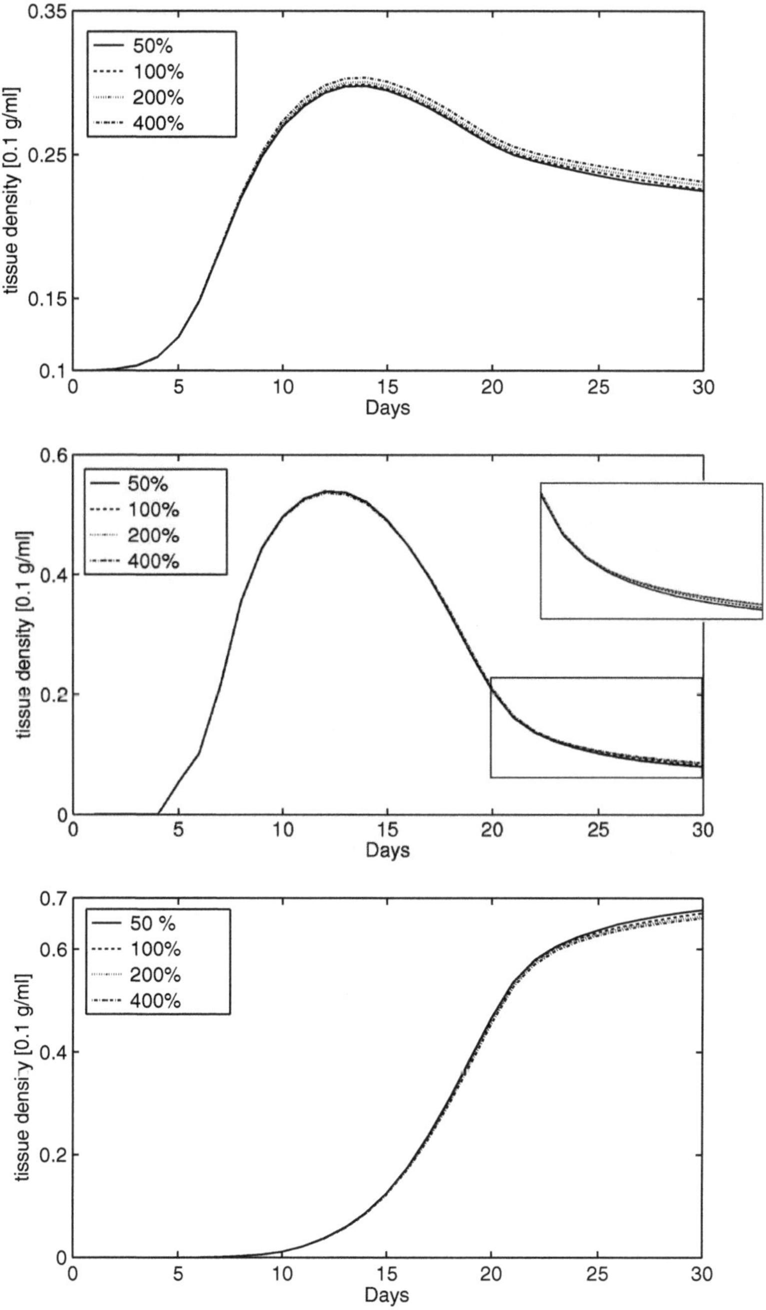

Fig. 6 Mean tissue densities in the callus resulting from the variation of stem cell differentiation to fibroblasts. *Top to bottom*: connective tissue, cartilage and immature bone

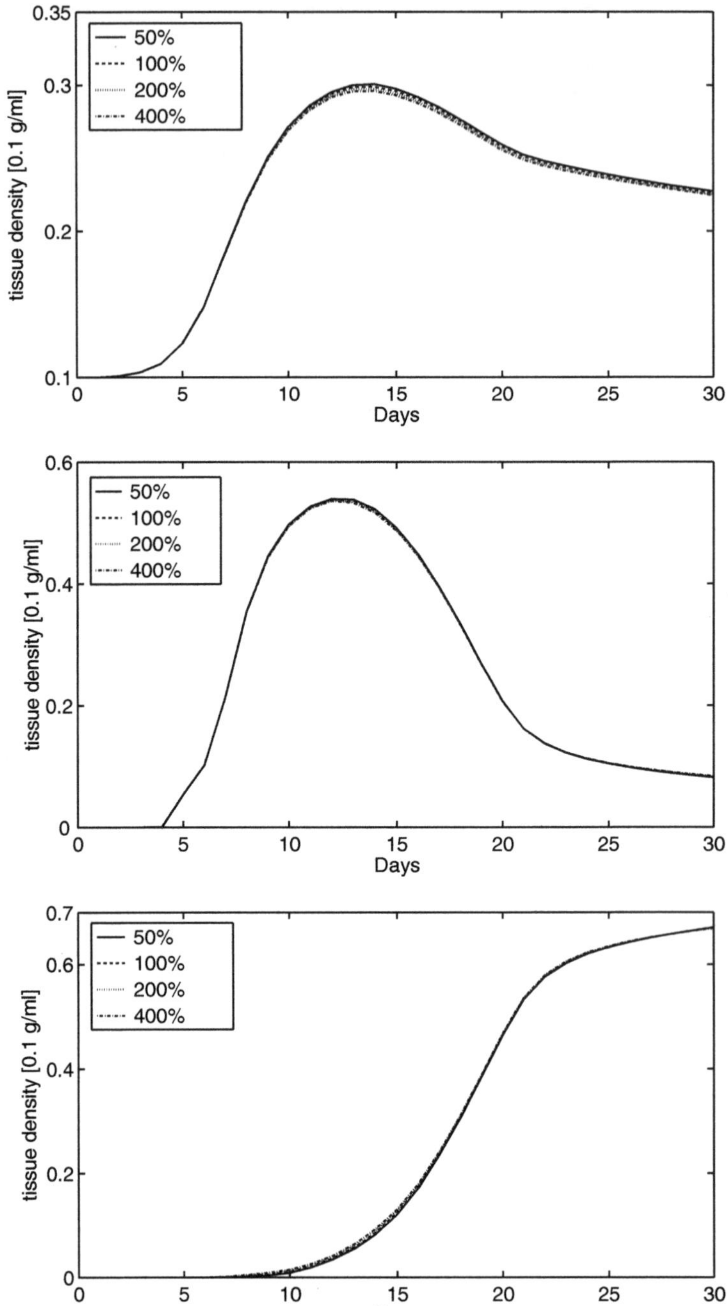

Fig. 7 Mean tissue densities in the callus resulting from the variation of stem cell differentiation to osteoblasts. *Top to bottom*: connective tissue, cartilage and immature bone

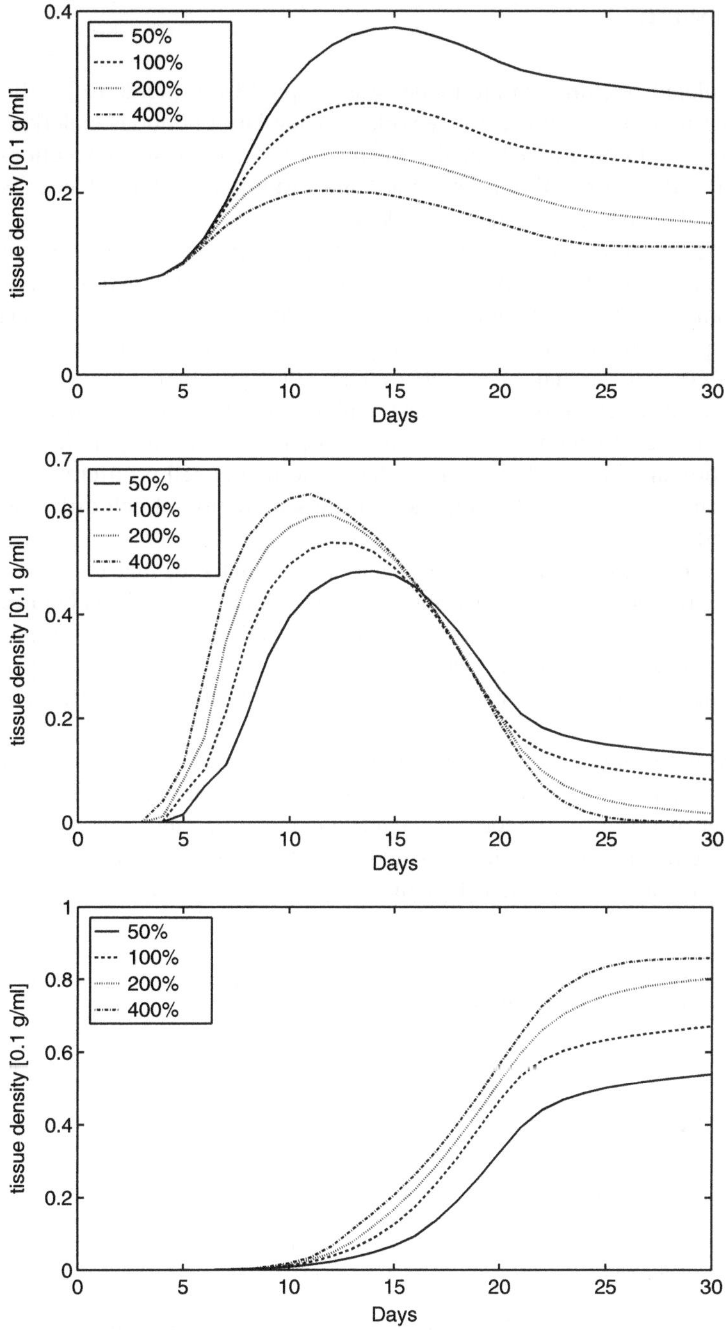

Fig. 8 Mean tissue densities in the callus resulting from the variation of stem cell differentiation to chondrocytes. *Top to bottom*: connective tissue, cartilage and immature bone

5 Discussion

A bioregulated fracture healing model was adapted from literature [7, 8] and implemented in a finite element framework. A Time Discontinuous Galerkin scheme was coupled with the Finite Calculus method [14] to obtain stable solutions for the advection-diffusion-reaction problem at hand. Additionally, a mechanical stimulus for stem cell differentiation has been added.

This model was able to predict the general healing pattern of a long bone fracture. The evolution of the tissue distribution is in good agreement with descriptions of the bone repair histology in literature. However, as seen in Fig. 5 the simulation predicts a very expeditious healing. The interfragmental movement decreases very rapidly from day 10 to 20, so that the bone has already then regained its functionality.

Stem cell differentiation to fibroblasts and osteoblasts has no significant influence on the tissue development for a wide range of parameters. The population of these cells is dominated by the logistic growth used to model cell proliferation. This has to be taken into account when the mechanical stimulation of cell differentiation is considered.

From a numerical point of view it has to be remarked, that the mathematical model incorporates a vast number of parameters, which rely on data obtained in very few studies. While this means the shown computations cannot safely predict the development of an individual healing fracture, an analysis of sensitive parameters helps to identify critical parameters of the biological entities taking part in the repair process.

Future work might include extending the mechanical stimulation to additional cell functions like proliferation, growth factor or matrix production and incorporating further measures of the local mechanical demand like the hydrostatic pressure. A 3D implementation of the model would allow for a more complex bone and callus geometry, i.e. an asymmetric callus and enable a greater variety of load cases like bending.

Acknowledgments This research was founded by the German Research Foundation (Deutsche Forschungsgemeinschaft Na330/8-1).

References

1. Carter, D.R., Blenman, P.R., Beauprè, G.S.: Correlations between mechanical stress history and tissue differentiation in initial fracture healing. J. Orthop. Res. **6**, 736–748 (1988)
2. Lacroix, D., Prendergast, P.J., Li, G., Marsh, D.: Biomechanical model to simulate tissue differentiation and bone regeneration: application to fracture healing. Med. Biol. Eng. Comput. **40**, 14–21 (2002)
3. Garcìa-Aznar, J.M., Kuiper, J.H., Gòmez-Benito, M.J., Doblarè, M., Richardson, J.B.: Computational simulation of fracture healing: influence of interfragmentary movement on the callus growth. J. Biomech. **40**, 1467–1476 (2007)

4. Claes, L.E., Heigele, C.A.: Magnitudes of local stress and strain along bony surfaces predict the course and type of fracture healing. J. Biomech. **32**, 255–266 (1999)
5. Witt, F., Petersen, A., Seidel, R., Vetter, A., Weinkammer, R., Duda, G.N.: Combined in vivo/in silico study of mechanobiological mechanisms during endochondral ossification in bone healing. Ann. Biomed. Eng. **39**, 2531–2541 (2011)
6. Chen, G., Niemeyer, F., Wehner, T., Simon, U., Schuetz, M.A., Pearcy, M.J., Claes, L.E.: Simulation of the nutrient supply in fracture healing. J. Biomech. **42**, 2575–2583 (2009)
7. Bailón-Plaza, A., van der Meulen, M.C.H.: A mathematical framework to study the effects of growth factor influences on fracture healing. J. Theor. Biol. **212**, 191–209 (2001)
8. Geris, L., Gerisch, A., Vander, Sloten J., Weiner, R., Van Oosterwyck, H.: Angiogenesis in bone fracture healing: a bioregulatory model. J. Theor. Biol. **251**, 137–158 (2008)
9. Bailón-Plaza, A., van der Meulen, M.C.H.: Beneficial effects of moderate, early loading and adverse effects of delyed or excessive loading on bone healing. J. Biomech. **36**, 1069–1077 (2003)
10. Geris, L., Van Vander Sloten, J., Oosterwyck, H.: Connecting biology and mechanics in fracture healing: an integrated mathematical modeling framework for the study of nonunions. Biomech. Model. Mechanobio. **9**, 713–724 (2010)
11. Sapotnick, A., Nackenhorst, U.: A combined fic-tdg finite element approach for the numerical solution of coupled advectiondiffusionreaction equations with application to a bioregulatory model for bone fracture healing. Int. J. Numer. Meth. Eng. **92**, 301–317 (2012)
12. Duda, G.N., Eckert-Hübner, K., Sokiranski, R., Kreutner, A., Miller, R., Claes, L.: Analysis of inter-fragmentary movement as a function of musculoskeletal loading conditions in sheep. J. Biomech. **31**, 201–210 (1998)
13. Oñate, E.: Derivation of stabilized equations for numerical solution of advective-diffusive transport and fluid flow problems. Comput. Methods Appl. Mech. Engrg. **151**, 233–265 (1998)
14. Oñate, E., Miquel, J., Hauke, G.: Stabilized formulation for the advection-diffusion-absorption equation using finite calculus and linear finite elements. Comput. Methods Appl. Mech. Engrg. **195**, 3926–3946 (2006)
15. Street, J., Winter, D., Wang, J.H., Wakai, A., McGuinness, A., Redmond, H.P.: Is human fracture hematoma inherently angiogenic? Clin. Orthop. Relat. Res. **378**, 224–237 (2000)

The Customized Artificial Hip Cup: Design and Manufacturing of an Innovative Prosthesis

Stefanie Betancur Escobar, Anas Bouguecha, Amer Almohallami, Henning Niemeier, Karin Lucas, Christina Stukenborg-Colsman, Ingo Nolte, Patrick Wefstaedt and Bernd-Arno Behrens

Abstract The demand for customized products has increased in recent years and will become even more important in the future. This trend is mainly observed for the medical technology sector influenced by the increasing manufacturing of patient-individual prostheses. In particular the manufacturing of customized hip cups is gaining in importance. Over 800,000 total hip replacements are performed world-wide each year. Despite this experience, the migration and loosening of the hip prosthesis especially of the cup due to the bone resorption caused by stress shielding is a current problem. Patient-specific hip cups can be used to counteract this. However, individual hip cups are only implanted for the treatment of great deformations or tumours because of the cost-intensive manufacturing. Within this project the

S. Betancur Escobar (✉) · A. Bouguecha · A. Almohallami · H. Niemeier · B. Behrens
Institute of Forming Technology and Machines, Leibniz Universität Hannover, Hannover, Germany
e-mail: betancur@ifum.uni-hannover.de

A. Bouguecha
e-mail: bouguecha@ifum.uni-hannover.de

A. Almohallami
e-mail: almohallami@ifum.uni-hannover.de

H. Niemeier
e-mail: niemeier@ifum.uni-hannover.de

B. Behrens
e-mail: behrens@ifum.uni-hannover.de

K. Lucas · I. Nolte · P. Wefstaedt
Small Animal Clinic, Foundation University of Veterinary Medicine Hannover, Hannover, Germany
e-mail: karin.lucas@tiho-hannover.de

I. Nolte
e-mail: ingo.nolte@tiho-hannover.de

P. Wefstaedt
e-mail: patrick.wefstaedt@tiho-hannover.de

C. Stukenborg-Colsman
Department of Orthopaedic Surgery, Hannover Medical School, Hannover, Germany
e-mail: christina.stukenborg@ddh-gruppe.de; matthias.lerch@ddh-gruppe.de

T. Lenarz and P. Wriggers (eds.), *Biomedical Technology*, Lecture Notes in Applied and Computational Mechanics 74, DOI 10.1007/978-3-319-10981-7_4

remodelling process is calculated with a conventional prosthesis via finite element method (FEM) coupled with multi-body simulation (MBS). A migration of the cup in the proximal direction can be suggested. Based on these results an innovative and economic concept for the design and production of patient-individual hip cups for primary surgery by means of sheet metal forming is developed. In this two-stage process first standardized titanium sheet metal components are produced. Then a true-size enlargement of these components is executed by a modified adaptive rubber-die forming process. The development is accompanied by an FE simulation-based planning as well as a metal forming adapted design method. In this study the first part of the design method is demonstrated, which contains the deduction of a universal acetabular geometry, necessary for the production of the standardized component. Furthermore, high pressure sheet metal forming (HPF) will be introduced for the manufacturing of standardized components. Therefore an FE-simulation of the process is carried out for the design of the forming tool.

Keywords Total hip arthroplasty · Agglomerative clustering · Sheet metal forming · Finite-element-simulation

1 Introduction

One of the most common operations in Germany is the total hip arthroplasty (THA) with more than 200,000 implantations each year [1]. Over 800,000 are even performed in the whole world [2]. Considering the steadily growing life expectancy of the patients, the prostheses have to remain in the body for ever increasing periods of time. Despite of years of experience and good clinical results, different complications occur which negatively influence the retention time of the implants in the body. Up to 7 % of all implanted prostheses will be replaced after 10 years as a consequence of the aseptic loosening of one implant component [3]. Especially the migration or loosening of the artificial hip cup due to bone remodelling is still a current problem [4, 5]. With regard to the ageing population this means an increase in revision surgeries marked with a higher risk of complications and thus increasing cost for the health care system.

The migration or loosening may be a result of the changed mechanical conditions after the implantation of the THA. This change can lead to a bone remodelling caused by stress shielding [6]. Furthermore, for a stable anchoring of the hip cup a high bone resection is necessary which promotes bone resorption and hampers revision surgery. A patient-specific solution can be used to counteract these disadvantages of conventional hip cups. However, individual hip cups are only implanted for the treatment of great deformations or tumours. This is due to the time- consuming and cost-intensive production of each components [7].

The overall aim of this project is the development of an innovative and economic concept for the design and production of patient-individual hip cups for primary surgery by means of sheet metal forming. Specific limitations (e.g. rigid tools) of conventional forming processes normally prevent a time-saving and economically reasonable patient-individual production. Nevertheless, the use of metal forming for the manufacturing of customized products is necessary because of excellent mechanical component properties, extensive shaping formabilities as well as optimal material utilization. Currently potential approaches to an individual production are unitized tools, CNC-controlled incremental forming methods or multi-point forming [8–10].

The manufacturing concept developed in this project consists of an innovative two-stage sheet metal forming process. The development is accompanied by a FE simulation-based planning as well as a and a new design method adjusted to the process. The following sections of this paper give the results of a numerical investigation of the bone remodelling process with a conventional prosthesis via finite element method (FEM) coupled with multi-body simulation (MBS) as well as the idea of the concept for the manufacturing of the individual hip-cup. Based on already executed comparative simulations, the high pressure sheet metal forming (HPF) is introduced for the manufacturing of the standardized components. Afterwards the first part of the design method is demonstrated, which contains the deduction of a universal acetabular geometry, necessary for the production of the standardized component.

2 Bone Remodelling

As already stated, the migration of hip-cup prostheses due to bone remodelling is still a problem. A numerical method for the prediction of the bone remodelling after implantation of a artificial hip-cup is utilized below to emphasize the need of customized prostheses.

By using a THA the physiological load distribution in the femur and the acetabulum is modified. This modification is due to the fact that the bone stiffness is considerably lower than the stiffness of the implant material. Thus, a large proportion of the load is not directed over the bone but preferably over the prosthesis. In accordance with Wollfs law, which says that changes in form and function of a bone lead to changes in internal structure and external form [11], the so-called strain-adaptive bone remodelling occurs, which can lead to a migration or loosening of the prosthesis [12, 13].

Due to the mentioned problems a qualitative and quantitative prediction of the bone remodelling is important for implant design. At the Institute of Forming Technology and Machines a simulation method was developed to calculate the strain-adaptive bone remodelling in the periprosthetic femur via FEM [14, 15]. Furthermore, the bone remodelling in the pelvis after total hip replacement was numerically

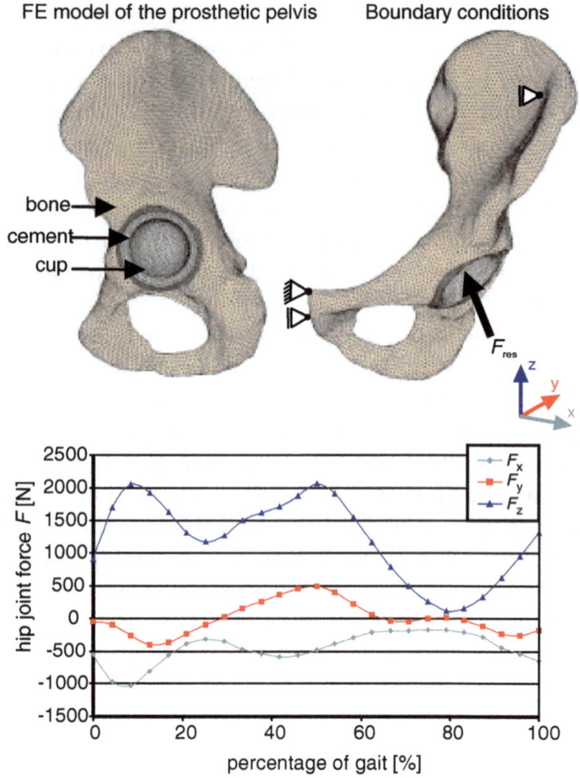

Fig. 1 FE-model of the periprosthetic pelvis and components of the calculated resultant hip joint forces [17]

simulated on the basis of the established simulation method [16, 17]. In Fig. 1 the FE-model of this simulation is illustrated, consisting of the bony structure provided with a cemented cup. In addition to the FE-model the components of the calculated resultant hip joint forces used as boundary conditions are pictured in Fig. 1. For the determination of hip joint forces an MBS model of a human test person with a normal walking speed (1.1 m/s) was used.

According to the simulation with load conditions derived from the MBS a total bone loss of 1.4 % can be assumed in the periprosthetic pelvis. The density distributions in the periprosthetic pelvis for the initial and final state are illustrated in Fig. 2. A high bone resorption can be observed in the acetabulum. On the basis of the final density distribution in the acetabulum a migration of the conventional cup can be suggested in proximal direction.

This migration can be avoided by means of patient-individual hip cups. In the following section a new concept for the design and production of patient-individual hip-cups prostheses out of titanium sheets will be introduced and the results of the concept development will be presented.

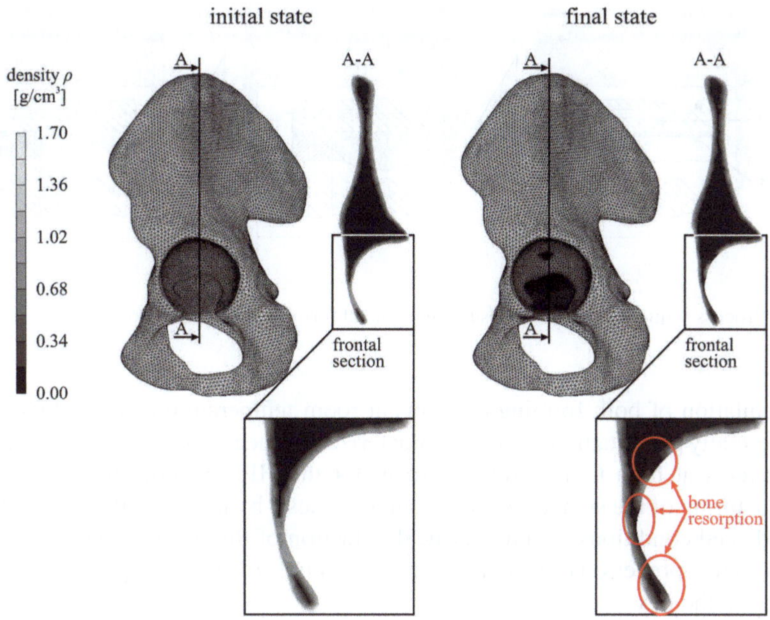

Fig. 2 Density distributions in the periprosthetic pelvis for the initial and final state [17]

3 Manufacturing Concept

It is planned to produce the patient-individual hip-cup prosthesis by a two-stage manufacturing process consisting of the following steps: The first step is the production of standardized titanium sheet metal components with undersize. These standardized components can be produced by means of punch-die forming ("conventional" deep drawing with punch, blank holder and die) or HPF in large-scale production. In the second step the true-to-size enlargement of the produced standardized components is executed. This enlargement is realized by double acting rubber-die forming with adjustable elastomer lower die considering the present patient-individual geometries of the acetabula.

In the following the production of the standardized titanium sheet metal components is considered in detail. Thereby comparative simulations are presented for the identification of the appropriate method for the first production step. Taking into account these simulations, the HPF-process will be introduced to the production of the standardized components.

3.1 Comparative Simulations

Comparative simulations were executed to identify the most convenient method to produce the standardized prosthesis [18]. This preliminary investigation consists of

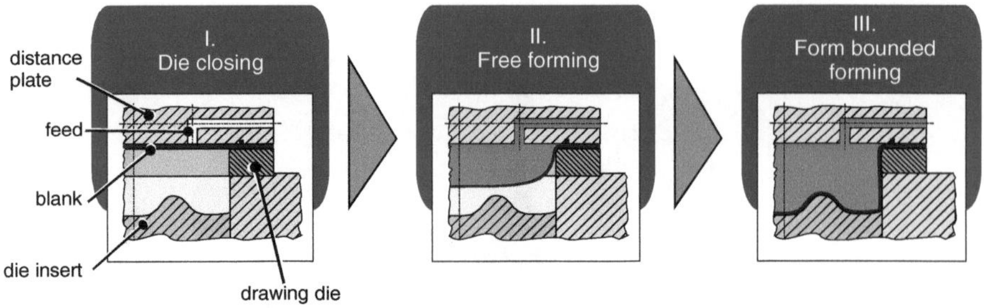

Fig. 3 Process sequence of high-pressure sheet metal forming with die insert

the simulation of both forming concepts at room temperature and at 200 °C using commercially pure titanium grade 2 characterized by means of standard tentile testing experiments at both temperatures. Except for the HPF at room temperature, the manufacturing of the standardized hip-cup prosthesis by means of the other forming methods and conditions lead to a critical reduction of the sheet thickness (>25 %). Thus the HPF-process at room temperature was choosen for the real process described in detail in the next section.

3.2 High Pressure Sheet Metal Forming

The HPF process offers, in addition, the following advantages over the punch-die forming procedure [19]:

- use of different blank thicknesses and qualities without changing the die,
- production with less stages,
- improvement of buckling strength as well as dimensional and geometrical accuracy.

In Fig. 3 the process sequence is illustrated. First the die is closed, so that blank and distance plate are in contact. This leads to a sealed interior space. Then the fluid is pumped into the interior space of the die through the feed in the distance plate. During the first forming phase a free forming take place. In the second phase, the form-bounded forming takes place, so that the blank is pressed against the contour of the die insert.

4 Derivation of the Universal Prosthesis Geometry

To realize the manufacturing concept the development of the two-stage manufacturing process is accompanied by a design method adjusted to the two-stage manufacturing process presented before. For the manufacturing of the standardized components, a universal acetabulum geometry should be derived from the CT-data of real hip joints. By means of reverse engineering an associative parametric model is generated based on this universal geometry. The parametric composition of the produced universal geometry finally allows the adjustment of the parametric model to the individual geometry of the patient. For this geometrical superposition and a parameter adaptation is planned. The resulting individualized geometry is used for the enlargement step with the double acting rubber-die forming.

In this section the approach to the generation of the universal prosthesis geometry will be introduced. Fist the general design chain of the generation of the universal geometry is presented as well as a preliminary investigation. Subsequently the agglomerative clustering is introduced and utilized for the reproducible deduction of the universal geometry.

4.1 General Design Chain

The method for the derivation of the universal prosthesis geometry is firstly implemented by means of canine pelvis geometries. In Fig. 4 the design chain for the deduction of the universal geometry is illustrated.

First of all, a multitude of CT examinations are executed to receive a high number of pelvis geometries. Afterwards the collected CT Data are segmented, so that only the geometries of the pelvis are present as 3D models. For this the Open Source Software YaDiV 1.0 beta4 (WelfenLab, Germany) is used. Then the 3D-models of the pelvis will be trimmed reproducible using the software Geomagic Studio 2012 (Geomagic, USA) so that only the relevant part of the acetabulum remains. These trimmed geometries are used for the derivation of the universal geometry, realized by

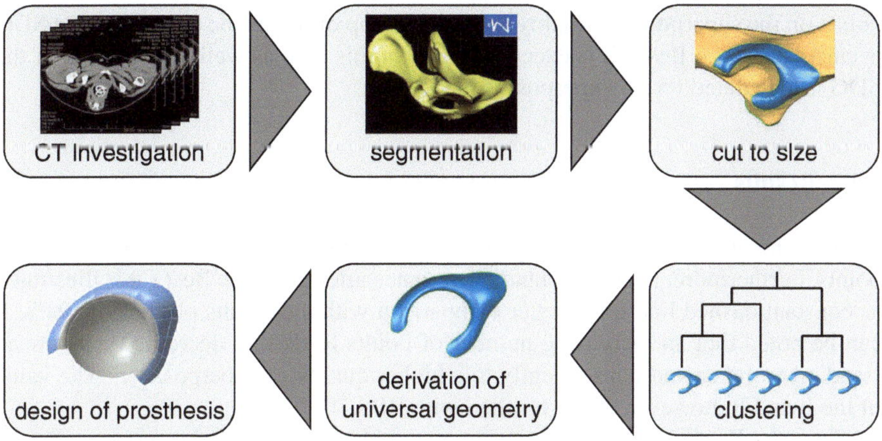

Fig. 4 Design chain for the derivation of the universal geometry [20]

the agglomerative clustering. This clustering method is described more precisely in the following. Finally an idealized hip cup prosthesis is designed using the resulting universal geometry of the edge of the acetabulum.

4.2 Preliminary Investigation

4.2.1 Method

The basis for the use of the deduction of the universal geometry is the superposition of the trimmed 3D models. Therefore a suitable method has to be found. Thus a preliminary investigation was executed [18]: In this study two different superposition methods, the so called Best Fit and the Reference Point System (RPS), were compared. Both of them are provided by the commercial software Geomagic Qualify 2012 (Geomagic, USA). The analyses are realized with 11 segmented pelvis geometries prepared as described before by the design chain.

The generated data are compared in the same way. First one geometry is selected randomly from the 11 segmented pelvis geometries used for both superposition methods. This geometry is defined as reference for the following analysis. The remaining 10 geometries are aligned as test with the reference by using both methods. In this context a geometry defined as test is allowed to move against the geometry defined as reference which is fixed. After each superposition the **s**tandard **d**eviation of the **g**eometrical deviation (SDG) is determined. By means of the SDG the superposition quality can be assessed, because this value implies the information about the deviation over all the geometries.

For the RPS method 21 points are located by hand on reproducible points of the acetabular area. The RPS method is performed seven times with a varying number of utilized points (6–21 points) in order to examine the influence of the count of points on the superposition quality. For each group of points the average of the SDG is calculated. The Best Fit is executed once. In this case as well, the average of the SDG is calculated for all superpositions.

4.2.2 Results

In Fig. 5 the mean SDG after using the RPS is plotted against the varying number of points. Furthermore, the mean standard variance after using the Best Fit is illustrated as constant dashed line for a better comparison with the results of after the RPS. It can be noted that an increasing number of points leads to a decrease of the mean standard variance and consequently to a higher quality of superposition. The value of the Best Fit, however, is not reached with RPS.

Thus, the Best Fit seems to be the more suitable method for the superposition of the pelvis geometries and is selected for the derivation of the universal geometry.

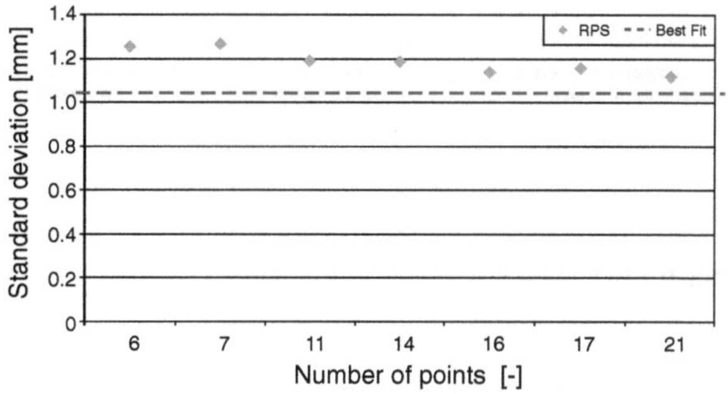

Fig. 5 Mean standard deviation of the geometrical deviation after the RPS plotted against the varying number of points [18]

Furthermore, the automatic, reliable and reproducible execution of this superposition method is a great advantage over the RPS.

4.3 Agglomerative Clustering

4.3.1 Method

The universal geometry is a geometry which maps all underlying original geometries. For the determination of this universal geometry, the original geometries should be reproducibly combined. After a comparison of different possible combination methods, agglomerative clustering was identified as the best method for the derivation of the universal acetabulum geometry.

Agglomerative clustering is subsumed under the so-called data mining method. Data mining is an approach to finding and describing existing but non-obvious knowledge in collected data. For this purpose data will be analyzed and screened as to conspicuousness, commonalities or generally to particularities [21, 22]. With means of agglomerative clustering similarities can thus be analyzed and different groups or outliers can be identified. Individual elements will be joined in groups due to a certain distance measure which is used as comparison criterion. The groups will be designated as cluster. The comparison criterion is defined depending on the application [23, 24].

4.3.2 Adaption of the Method

The adaption of agglomerative clustering to the derivation of the universal acetabulum geometry is illustrated in Fig. 6. Here it should be noted, that the clustering is realized only considering the geometrical features. Characteristics like body weight or breed are not taken into account. The whole procedure is developed on the basis of Geomagic Qualify.

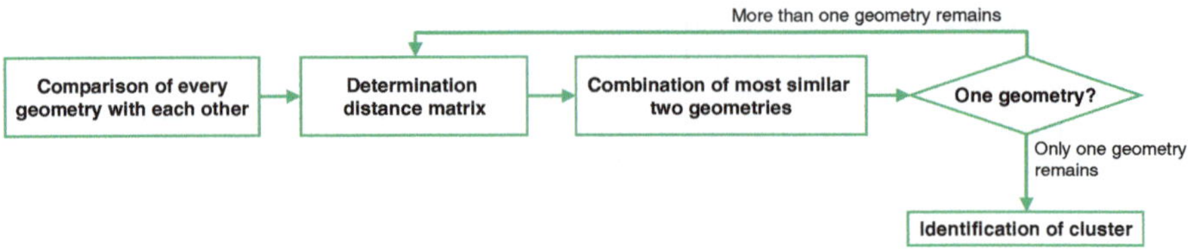

Fig. 6 Adapted agglomerative clustering

The input data for agglomerative clustering are the original acetabulum geometries prepared as described before. First of all, every original geometry is compared with each other. Thus two geometries are brought to superposition using Best Fit. Afterwards the SDG is determined. This value will be the comparison criterion in this case. In the second step a so-called distance matrix is written implying all determined SDG. This matrix is used to identify the two geometries with the lowest SDG used for the this step. Here the two most similar geometries are combined, realized by the derivation of a mean geometry.

The method for the creation of a mean geometry is as follows. After the identification of the most similar two geometries, a deviation table is exported which contains the coordinates of each point of reference as well as the test geometry and the deviation between the compared points. Based on this table the points of a new geometry are calculated by means of the following equation:

$$\begin{aligned} \vec{x}_{mean} &= \vec{x}_{ref} + \frac{\Delta\vec{x}}{2}, \\ \vec{x}_{mean}, \vec{x}_{ref}, \Delta\vec{x} &\in \Re^3 \end{aligned} \quad (1)$$

where $\vec{x}_{mean}$ is the mean point vector, $\vec{x}_{ref}$ is the reference point vector and $\Delta\vec{x}$ is deviation vector. After the creation of the mean geometry its original combined geometries will be removed from the geometry pool and the new mean geometry created in step three will replace these geometries. Then it will be tested if only one geometry remains or not. If not a new distance matrix is created with the new mean geometry. This loop is repeated until one geometry remains. Finally, based on this geometry different clusters are identified (see Fig. 6).

4.3.3 Derivation of a First Universal Geometry

By means of the described method, a clustering with 34 canine CT-data was carried out. The database only contains data of dogs which according to their weight ($\geq$15 kg) and anatomy would be suited to receiving a hip replacement. To identify various cluster the tree of combined pairs referred to as dendrogram is created.

In Fig. 7 the resulting tree of the analysis with 34 choosen data sets is illustrated. In this picture the geometries of the trimmed pelvis are represented by the green circle,

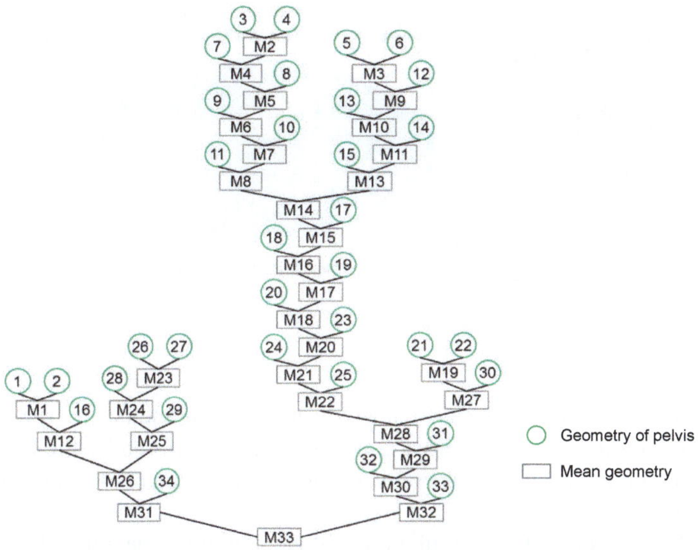

Fig. 7 Dendrogram after the clustering of 34 pelvis geometries

Fig. 8 Designed prosthesis by means of the derived universal geometry

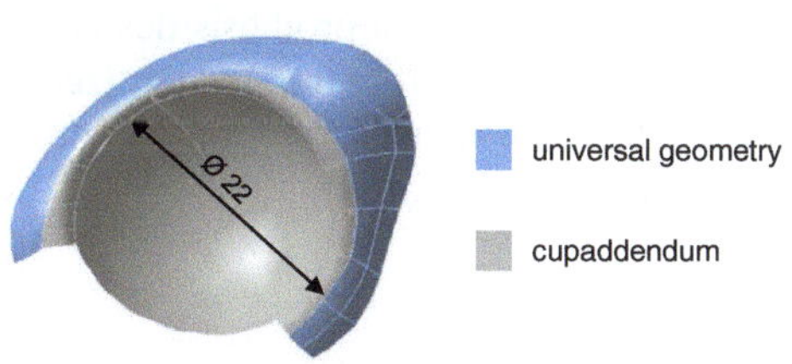

the mean geometries by the grey square. The numbering of the trimmed geometries and the mean geometries is executed in the order of combination and generation.

It can be noted that a minimum of 2 clusters results, these with the universal geometry $M31$ and with the geometry $M32$. For the design of a first universal hip cup and a first forming die the geometry $M32$ was used. This selection results the visual check of the dendrogram and from the fact that $M32$ represents the biggest amount of geometries.

4.3.4 Standardized Hip Cup Prosthesis

In Fig. 8 the derived hip cup prosthesis is pictured. This prosthesis consists of the deducted universal geometry of the trimmed pelvis (blue area) and an cup addendum (grey area).

The cup addendum is designed by means of Catia V5 (Dessault, France). For the design of those addendum a hemisphere with a diameter of 22 mm is created,

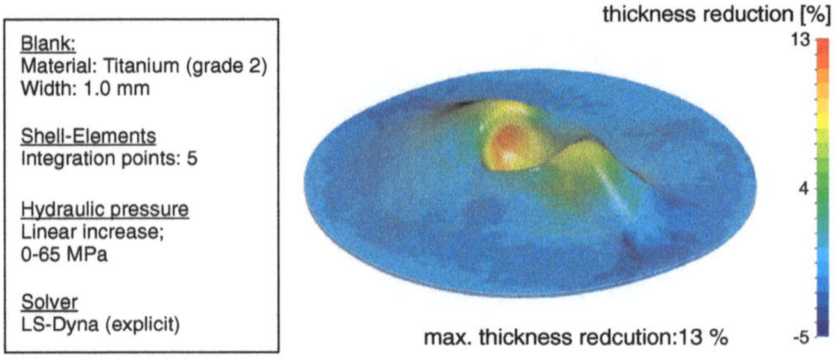

Fig. 9 Results of the process simulation [20]

which is trimmed and connected with the universal geometry. To verify that the designed standardized hip cup prosthesis can be manufactured by the HPF-process a FE-simulation is carried out. Thereby all settings are taken over by the simulations executed in [18].

In Fig. 9, the results of the simulation is illustrated. It can be noted that the manufacturing process lead to a reduction ratio of maximum 13 %. This value is even lower compared to the results presented in Sect. 3. This is due to the fact that for the comparative simulation an prothesis designed by an individual canine acetabulum was used which differs from the standardized prothesis. This verification leads to the use of the designed standardized hip-cup prosthesis for the transfer of the HPF-process to real process design.

5 Outlook

For the future it is planned to extend the presented geometrical deduction method, so that an automated identification of the cluster can be realized. Furthermore it is planned to execute a correlation hypothesis between the cluster formation and certain parameter. These parameter can be weight, length- and height-diameter of the acetabulum as well as acetabulum depth. On the basis of this analysis it can be possible to assign a potential patient to a cluster only by measuring the choosed parameter. Additionally an increase of the original geometry number is intended to provide a statistically secured data base.

Based on the simulations introduced in this study the design and the test of the HPF die for the production of the standardized components will be a next step of this project as well. With this die it is planned to produce first canin hip cups which will be tested. Cadaver test will be executed to provide information concerning the effort of the surgical intervention compared to the implantation of a conventional hip-cup prosthesis. Furthermore bone remodelling simulations using the new prosthesis design will be carried out. Finally it is planned to adapt the whole concept for human geometries.

Acknowledgments The study is carried out in the framework of the Collaborative Research Center 599 "Sustainable degradable and permanent implants out of metallic and ceramic materials" and is a part of the subproject D13 "Development, metal-forming and evaluation of patient-individual acetabular components". The authors would like to thank the German Research Foundation (DFG) for the financial support.

References

1. Schnabel, P., Borelli, S.: Endoprothesenregister: Höchsten Anforderungen genügen. Dtsch Arztebl **18**(48), A 2598–A 2602 (2011)
2. Fraldi, M., Esposito, L., Perrella, G., Cutolo, A., Cowin, S.C.: Topological optimization in hip prosthesis design. Biomech. Model. Mechanobiol. **9**, 389–402 (2010)
3. Garellick, G., Karrholm, J., Rogmark, C., Rolfson, O., Herberts, P.: Annual report 2011: Swedish hip arthroplasty register (2012)
4. Garcia-Cimbrelo, E., Diaz-Martin, A., Madero, R., Munera, L.: Loosening of the cup after low-friction arthroplasty in patients with acetabular protrusion: The importance of the position of the cup. J. Bone Joint Surg. Br. **82**, 108–115 (2000)
5. Laursen, M.B., Nielsen, P.T., Søballe, K.: Bone remodelling around ha-coated acetabular cups: A dexa study with a 3-year follow-up in a randomised trial. Int. Orthop. **31**, 199–204 (2007)
6. Behrens, B.A., (ed).: Umformtechnik - Ein Wirtschaftszweig mit Potential: 19. Umformtechnisches Kolloquium Hannover [UKH], 27. und 28. Februar 2008, Hannover (2008)
7. Zäh, M., Hagemann, F.: Wirtschaftliche Fertigung mit Rapid-Technologien: Anwender-Leitfaden zur Auswahl geeigneter Verfahren. Kostengünstig produzieren. Hanser, München (2006)
8. Hirt, G., Ziegler, S., Bambach, M.: Recent developments in incremental sheet forming for prototype and small series. ymposium "Current State and Future Trends of Metal Forming Technology in Automobile Industries", 27.02.2008, Nagoya, Japan (2008) 69–80.
9. Meichsner, T.P.: Tool-in-tool Innovative und wandlungsfähige Großwerkzeuge für den Karosseriebau. In Behrens, B.A., ed.: Umformtechnik - Ein Wirtschaftszweig mit Potential, Hannover (2008) 129–136.
10. Tan, F.X., Li, M.Z., Cai, Z.Y.: Research on the process of multi-point forming for the customized titanium alloy cranial prosthesis. 3rd International Conference on Advanced Forming and Die Manufacturing Technology 187–188 (2007) 453–457.
11. Wolff, J.: Das Gesetz der Transformation der Knochen. 1. aufl., reprint [der ausg.] berlin, hirschwald, 1892 / hrsg.: Georg bergmann und georg duda edn. Pro Business, Berlin (2010).
12. Engh, C.A., McGovern, T.F., Bobyn, J.D., Harris, W.H.: A quantitative evaluation of periprosthetic bone-remodeling after cementless total hip arthroplasty. The Journal of bone and joint surgery. American volume 74 (1992) 1009–1020.
13. Kerner, J., Huiskes, R., van Lenthe, G.H., Weinans, H., van Rietbergen, B., Engh, C.A., Amis, A.A.: Correlation between pre-operative periprosthetic bone density and post-operative bone loss in tha can be explained by strain-adaptive remodelling. J. Biomech. **32**, 695–703 (1999)
14. Bouguecha, A.: Numerische und experimentelle Untersuchungen zum beanspruchsadaptiven Knochenumbau im periprotheischen caninen Femur. Volume 2007, Bd. 04 of Berichte aus dem IFUM. PZH, Produktionstechn. Zentrum, Garbsen (2007).
15. Behrens, B.A., Nolte, I., Wefstaedt, P., Stukenborg-Colsman, C., Bouguecha, A.: Numerical investigations on the strain-adaptive bone remodelling in the periprosthetic femur: Influence of the boundary conditions. BioMedical Engineering OnLine **8**, 1–9 (2009)
16. Bouguecha, A., Elgaly, I., Stukenborg-Colsman, C., Lerch, M., Nolte, I., Wefstaedt, P., Matthias, T., Behrens, B.A.: Numerical investigations of the strain-adaptive bone remodeling in the prosthetic pelvis. In: Magjarevic, R., Bamidis, P.D., Pallikarakis, N. (eds.) XII Mediterranean Conference on Medical and Biological Engineering and Computing 2010. IFMBE Proceedings, vol. 29, pp. 562–565. Springer, Heidelberg (2010)

17. Bouguecha, A., Weigel, N., Betancur Escobar, S., Nolte, I., Wefsteadt, P., Stukenborg-Colsman, C., Behrens, B.A.: Influence of assumed boundary conditions derived from mbs on numerically simulated strain-adaptive bone remodeling in the pelvis after total hip replacement. In: Proceedings of the 50th EMBEC (2011)
18. Behrens, B.A., Escobar, S.B., Almohallami, A., Weigel, N., Vucetic, M., Stukenborg-Colsman, C., Lerch, M., Nolte, I., Lucas, K., Wefstaedt, P., Bouguecha, A.: Production of patient-individual hip cups by sheet metal forming: simulation-based planning and metal forming adapted design method. Adv. Mater. Res. **907**, 253–264 (2014)
19. Doege, E., Behrens, B.A.: Handbuch Umformtechnik: Grundlagen, Technologien, Maschinen. VDI-Buch. Springer, Heidelberg (2010)
20. Betancur Escobar, S., Bouguecha, A., Almohallami, A., Niemeier, H., Nolte, I., Lucas, K., Stukenborg-Colsmann, C., Lerch, M., Behrens, B.A.: Patient-individual hip cups: Simulation-based design and sheet metal forming manufacturing. Biomedizinische Technik. Biomedical engineering (2013)
21. Ester, M., Sander, J.: Knowledge Discovery in Databases: Techniken und Anwendungen. Springer, Berlin (2000)
22. Herrmann, H.: Data Mining. 1. aufl edn. Grin Verl, [München [u.a.]] (2004)
23. Bacher, J.: Clusteranalyse: Anwendungsorientierte Einführung. 2, erg. aufl edn. Oldenbourg, München and Wien (1996)
24. Backhaus, K.: Multivariate Analysemethoden: Eine anwendungsorientierte Einführung; mit 6 Tabellen. 11, überarb. aufl edn. Springer-Lehrbuch. Springer, Berlin [u.a.] (2006)

On the Role of Phase Change in Modelling Drug-Eluting Stents

Franz Bozsak, Jean-Marc Chomaz, Abdul I. Barakat and Giuseppe Pontrelli

Abstract A model of drug release from an eluting stent to the arterial wall is presented. The coating layer is described as a porous reservoir where the drug is initially loaded in a polymer-encapsulated solid phase, and is then released both to the coating and to the tissue of the arterial wall in a free phase. The wall is treated as a heterogeneous porous medium and the drug transfer through it is modeled by a non-homogeneous set of coupled partial differential equations that describe a convection-diffusion-reaction process. Change of phases due to drug dissolution in the coating and binding-unbinding reactions in the arterial wall are addressed. Numerical results show a strong coupling of the release kinetics in the polymer and the drug dynamics in the wall, and this coupling depends on the physico-chemical drug properties, the microstructure of the polymeric stent coating and the properties of the arterial wall.

Keywords Drug delivery · Drug-eluting stents · Local mass non-equilibrium · Two-phase mass transfer · Diffusion-convection-reaction equations

1 Introduction

Drug-eluting stents (DES) have drastically reduced the rate of restenosis compared to bare-metal stents and have since become the most common choice for the treatment of coronary arteries afflicted with advanced atherosclerotic lesions. DES consist of a metallic wire mesh platform coated with a polymer film that encapsulates a therapeutic drug aimed at preventing hyperplasia of smooth muscle cells (SMCs) responsible for the re-occulsion of the treated artery, termed restenosis. To ensure effective performance, both the stent geometry and the coating design need to be optimized. The success of an antiproliferative drug therapy from a DES depends on

F. Bozsak · J.-M. Chomaz · A.I. Barakat
Laboratoire D'Hydrodynamique, Ecole Polytechnique, CNRS UMR7646, Palaiseau, France

G. Pontrelli (✉)
IAC - CNR, Via Dei Taurini 19, Rome, Italy
e-mail: giuseppe.pontrelli@gmail.com

T. Lenarz and P. Wriggers (eds.), *Biomedical Technology*, Lecture Notes in Applied and Computational Mechanics 74, DOI 10.1007/978-3-319-10981-7_5

the amount of drug eluted from the stent, the rate of drug release, accumulation of drug and drug binding to cells in the arterial wall [1]. The local drug concentrations achieved are directly correlated with the biological effects and local toxicity, and establishing the optimum dose to be delivered to the tissue remains a challenge in today's DES design and manufacturing [2, 3]. The two most common drugs used today are paclitaxel and derivatives of rapamycin (the so called limus drugs).

Although many studies of DES efficacy and optimal design have been carried out using either experimental methods [4] or numerical simulations [5, 6], many questions remain unanswered. Validated mathematical models for computing the drug concentration in the arterial wall can provide a useful tool in the manufacturing and development of new and more efficacious DES [7]. Such models should incorporate the pharmacokinetics responsible for the drug release to study the effect of different coating parameters and configurations on drug elution [8]. Hossainy and Prabhu developed a mathematical model to predict the transport reaction of drug release in biodurable materials and biodegradable polymers [9]. Although the polymer acts as the drug reservoir and a strategic design of its characteristics would improve the release performance, studies probing the drug elution process from the coating platform are limited. In most studies, the coating is considered as a continuum where the drug is incorporated directly into the liquid phase. However, at the microscopic scale, the polymer is a porous medium where the solid and fluid phases coexist [10]. In particular, the solid matrix acts as a drug reservoir, where the drug is initially bound to the solid phase. Subsequently, after stent insertion, expansion and contact with vascular tissues, a part of the drug is first transferred to the fluid phase, at a rate that depends on the porosity, permeability, and drug characteristics, and it then diffuses into the surrounding tissues.

Drug transport depends on the properties of the "coating-wall" system, taken as a whole and modeled as a coupled two-layered system. The multiphase release of drug from the coated stent and its distribution in the arterial wall must be carefully tailored to achieve the optimal therapeutic effect and to deliver the correct dose in the required time [11, 12]. The pharmacological effects of the drug as well as its tissue accumulation, duration and distribution could potentially have an effect on the drug's efficacy, and a delicate balance between adequate amount of drug delivered over an extended period of time and minimal local toxicity needs to be struck [13]. Thus, the model also needs to properly describe the drug dynamics in the different layers of the arterial wall. Although a large number of mathematical models are available to describe drug transport in arterial tissue, only a few [14, 15] consider the dynamic nature of the interaction of the drug with the cells of the wall. The released drug targets and binds to specific receptors on the surface of SMCs to block the uncontrolled proliferation and migration of these cells. Similar to the process in the coating, this entails a phase change of the mobile drug, which is transported through the interstitial space of the arterial wall, to a state where the drug is bound to the surface of the SMCs to exert its therapeutic effect.

In the present work, we model the coupled coating-wall system: we investigate the effect of phase change in both layers (the coating and the wall) by combining previous models, where a multi-layered porous wall model has been proposed and the influence of drug dissolution in the coating has been addressed [16, 17] with a model where the effect of different modeling assumptions on the predictions of the transport of sirolimus and paclitaxel in stented arteries has been studied [15]. All other mechanical effects (such as compression and expansion) due to the metallic or polymeric degradation/erosion are neglected.

Our results demonstrate a strong coupling of the release kinetics in the polymer and the drug dynamics in the wall. The results highlight the importance of tailoring drug release to drug kinetics in the arterial wall to ensure optimal DES performance.

2 A Two-Layer Model for Drug Elution

We consider a stent that is coated with a thin layer (of thickness l_0) of a porous polymer containing a drug and that is embedded into the arterial wall. As the bulk of drug transport occurs along the direction normal to the polymeric surface (radial direction) and by assuming axial symmetry, we can restrict our study to a simplified one-dimensional model (Fig. 1). In particular, we consider a radial line crossing the metallic strut, the coating and the arterial wall and pointing outwards. Because the wall thickness is very small compared to the radius of the artery, a Cartesian coordinate system x is used along the radial line. For simplicity, we consider the situation where the stent polymer is in direct contact with the medial layer of the arterial wall (hereafter simply referred to as *wall*). This layer is modeled as a homogeneous porous medium of thickness l_1. Without loss of generality, we assume $x = 0$ is the coating-wall interface (Fig. 1).

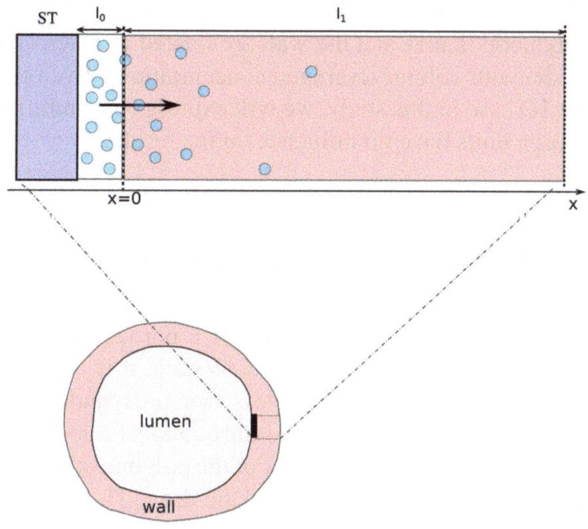

Fig. 1 Cross-section of the stent mesh, the coating, and the arterial wall: geometrical configuration and reference system (figure not to scale)

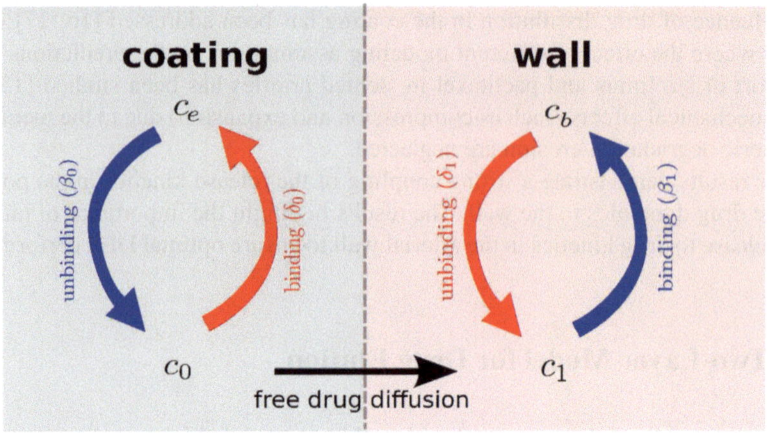

Fig. 2 Schematic of the cascade mechanism of drug delivery in the coating-wall coupled system. An unbinding (resp. binding) reaction occurs in the coating (resp. in the wall) (*blue arrows*). In both layers, reverse reactions (*red arrows*) are present in a dynamic equilibrium. Drug transport occurs only in the free phases c_0 and c_1

In this paper we are interested in non-local mass transfer processes in the coating, where the drug passes from a solid (polymer-encapsulated, c_e) to a free phase (c_0) by dissolution. Similarly, a part of the dissolved and transported drug in the wall (c_1) is metabolized by the cells and transformed to a bound state (c_b). Thus, the drug delivery process starts in the coating and ends at the SMC receptors, with bidirectional phase changes in a cascaded sequence, as schematically depicted in Fig. 2. A microscopic approach would require knowledge of the specific and local geometry of the individual pore structure networks, which is unfeasible. Therefore, both the polymeric matrix and the wall are treated as macroscopically homogeneous porous media with volume-averaged concentrations. Even though we are only considering the 1D case in this study, we will express concentrations in units of mol m^{-3}. These chosen units have no influence on the results.

2.1 The Two-Phase Coating Model

The coating of a DES consists of a porous polymeric matrix that encapsulates a therapeutic drug in solid phase; as such, it is unable to diffuse and to be delivered into the tissue [5]. Nevertheless, when expanded and deployed into the wall, the stent coating is exposed to the surrounding biological fluids. As a consequence, such fluids fill the interstitial spaces of the polymer and form a network of liquid channels, acting as a release medium for the drug. Thus, a fraction of the drug mass is first transferred, in a finite time, to the liquid phase, and then released and diffuses into the arterial wall. We carry out a mesoscale description of the volume-averaged drug concentrations in the liquid (c_0) and solid (c_e) phases considered separately. By using a microstructure argument [17], the governing equations are:

$$\frac{\partial c_e}{\partial t} = -\beta_0 c_e + \delta_0 c_0 \qquad \text{in } (-l_0, 0) \tag{1}$$

$$\frac{\partial c_0}{\partial t} = D_0 \frac{\partial^2 c_0}{\partial x^2} + \beta_0 c_e - \delta_0 c_0 \qquad \text{in } (-l_0, 0) \tag{2}$$

where D_0 (m^2 s^{-1}) is the diffusion coefficient of the unbound solute, $\beta_0 \geq 0$ (s^{-1}) and $\delta_0 \geq 0$ (s^{-1}) are the unbinding (*dissociation*) and binding (*re-association*) rate constants in the coating, respectively [10]. It should be noted that β_0 depends on the porosity of the coating ε_0 [17]. The binding rate constant is defined as the inverse of the characteristic solid-liquid transfer time scale, $\delta_0 = t_0^{-1}$. The ratio of the unbinding and binding rate constants is the equilibrium dissociation constant $K_0 = \dfrac{\delta_0}{\beta_0} = \dfrac{1-\varepsilon_0}{\varepsilon_0}$. t_0 and K_0 are quantities that can typically be determined experimentally.

The associated initial conditions are:

$$c_e(x, 0) = C_e \qquad c_0(x, 0) = 0 \tag{3}$$

expressing that, at initial time, the entire drug exists in the solid phase at a maximum constant concentration, and it is subsequently released into the liquid phase. Since the metallic strut is impermeable to the drug, no mass flux passes through the boundary surface $x = -l_0$; hence, we impose a no-flux condition:

$$D_0 \frac{\partial c_0}{\partial x} = 0 \qquad \text{at } x = -l_0 \tag{4}$$

2.2 The Two-Phase Wall Model

Drug that enters the arterial wall is transported by convection and diffusion through the tortuous paths of the extracellular matrix surrounding the wall's cells. Similarly to the coating, a phase change can occur in the wall from the free state (c_1) of the drug to the bound state (c_b) and vice versa at the surface of the cells within the wall. The free drug (c_1) binds to specific receptors on the surface of SMCs to form a bound complex (c_b). The maximum density of specific receptors available to the drug is denoted as c_b^{max}. The formed drug-receptor complex is not permanent and can be dissolved after a typical time scale t_1. These two processes (binding and unbinding) are modeled by a second order reversible saturating binding equations [1]:

$$\frac{\partial c_1}{\partial t} = D_1 \frac{\partial^2 c_1}{\partial x^2} - V_1 \frac{\partial c_1}{\partial x} - \beta_1 c_1 \left(c_b^{max} - c_b\right) + \delta_1 c_b \qquad \text{in } (0, l_1) \qquad (5)$$

$$\frac{\partial c_b}{\partial t} = \beta_1 c_1 \left(c_b^{max} - c_b\right) - \delta_1 c_b \qquad \text{in } (0, l_1) \qquad (6)$$

where D_1 is the effective diffusivity of unbound drug, V_1 (m s^{-1}) is a convection velocity, and $\beta_1 \geq 0$ (mol^{-1} m^3 s^{-1}) and $\delta_1 \geq 0$ (s^{-1}) are the binding and unbinding rate constants, respectively [15]. The unbinding rate constant is defined by the inverse of the characteristic unbinding time scale, $\delta_1 = t_1^{-1}$. It should be noted that due to the cross product of c_b and c_1 in the binding term of Eq. (6), the units of β_1 are different from those of β_0 (s^{-1}). β_1 is related to δ_1 by the equilibrium dissociation constant $K_1 = \dfrac{\delta_1}{\beta_1}$ (mol m^{-3}). The initial conditions are:

$$c_1(x, 0) = 0 \qquad c_b(x, 0) = 0 \qquad (7)$$

Finally, a perfectly absorbing boundary condition is imposed at the wall limit:

$$c_1(x, 0) = 0 \qquad \text{at } x = l_1 \qquad (8)$$

At the coating-wall interface, we impose the balance of flux

$$D_0 \frac{\partial c_0}{\partial x} = D_1 \frac{\partial c_1}{\partial x} - V_1 c_1 \qquad \text{at } x = 0 \qquad (9)$$

and continuity of the liquid-phase concentration:

$$\frac{c_0}{k_0 \varepsilon_0} = \frac{c_1}{k_1 \varepsilon_1} \qquad \text{at } x = 0 \qquad (10)$$

where ε and k are the porosities and the partition coefficients in the respective layers.

2.3 Physiological Parameters

The two most common drugs used for DES are paclitaxel and sirolimus. These two drugs have similar molecular size and resulting transport properties (effective diffusivity and convection velocity), but drastically different binding and unbinding dynamics in the arterial wall, resulting in different binding β_1 and unbinding δ_1 rate constants and maximum binding site densities c_b^{max} [14]. Based on the parameters determined in [14, 15, 17], we have chosen a set of parameters that describes a generic drug that is similar in its transport properties to paclitaxel and sirolimus and lies in between paclitaxel and sirolimus for its binding/unbinding properties. The parameters of the reference model are summarized in Table 1. We consider the entire

Table 1 Model parameters of the reference model

Parameter	Coating (0)	Arterial wall (1)
ε	0.1	0.25
k	1	1
$D\ (\mathrm{m^2\,s^{-1}})$	1×10^{-14}	5×10^{-12}
$V\ (\mathrm{m\,s^{-1}})$	–	5×10^{-8}
$\delta\ (\mathrm{s^{-1}})$	1.2×10^{-6}	2.8×10^{-8}
K	9	$3 \times 10^{-3}\ (\mathrm{mol\,m^{-3}})$
$c_b^{max}\ (\mathrm{mol\,m^{-3}})$	–	0.2
$C_e\ (\mathrm{mol\,m^{-3}})$	100	–

polymer and wall tissue as accessible and thus $k_0 = k_1 = 1$. The values for δ_0 and δ_1 correspond to a characteristic solid-liquid transfer time $t_0 = 1$ day and an unbinding time scale $t_1 = 100\,\mathrm{h}$, respectively.

2.4 Numerical Simulation

The governing equations are discretized using a finite element method with second-order Lagrangian elements, implemented in the commercial software package COMSOL Multiphysics 4.3a (COMSOL AB, Burlington, MA, USA). The relative tolerance is set to 10^{-5} and the absolute tolerance to 10^{-10}. The time advancing scheme is a backward difference formulation with variable order and time step size. The domain is discretized by 2,000 equally spaced elements. Mesh independence of the solution was confirmed using a coarse, medium and fine meshes with 1,000, 2,000 and 4,000 elements, respectively, with a relative difference of less than 0.1 % between solutions of progressively more refined meshes.

3 Results and Discussion

We present the mean concentration of each phase of the eluted drug in each of the layers of the model. The time considered in our simulations was 4 weeks. Figure 3 illustrates the temporal evolution of the averaged concentrations for three characteristic values of the solid-fluid transfer time t_0. For the smallest transfer time of $t_0 = 0.1$ days, all drug is released from the solid phase in the polymer within 1 week (Fig. 3a) which leads to a spike in the fluid phase drug concentration at $\approx$2 days at an average concentration of $\approx$2 (Fig. 3b).

Slower release of the drug from the solid phase entails significantly lower peak concentrations (Fig. 3b). For $t_0 = 1$ day, the coating is almost entirely depleted within

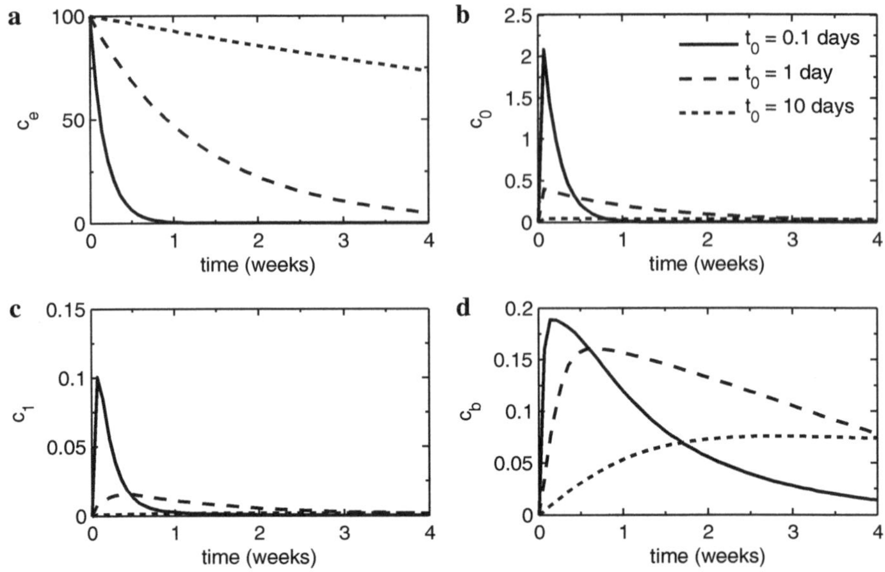

Fig. 3 Sensitivity analysis of the phase-change model in both layers to the characteristic solid-fluid transfer time scale t_0. Temporal evolution of **a** the solid-phase drug concentration c_e; **b** the liquid-phase drug concentration c_0; **c** the free drug concentration c_1; **d** the bound drug concentration c_b, each averaged over the entirety of their respective layers

the considered period (Fig. 3a). The slowest transfer time considered $t_0 = 10$ days causes only $\approx$75 % of the drug to be unbound from the polymer (Fig. 3a), leading to negligible liquid phase concentrations in the coating. We observe that the (excess) unbound concentration in the wall reaches significant levels only for the fastest solid-liquid transfer time (with a peak mean concentration of $\approx$0.1) (Fig. 3c). For $t_0 = 1$ day, the peak mean concentration reaches about $1/5$ of this value and is negligible for the slowest transfer time.

The dynamics of the uptake of the drug into the bound state in the wall are strongly coupled to the characteristic solid-fluid transfer time t_0 in the coating (Fig. 3d). For the fastest t_0, a high mean concentration of slightly less than 0.2 is reached within $\approx$1 day, implying that almost all receptors are occupied throughout the wall (compare to $c_b^{max} = 0.2$). With drug supply from the coating ceasing quickly, the retained concentration slowly decays over the next 4 weeks, reaching a mean concentration level of less than $1/10$ of the peak concentration at the end of the simulation. For slower transfer times, the period to reach peak concentration is longer and the magnitude of the peak concentration decreases. For $t_0 = 1$ day, the peak concentration is reached at $\approx$5 days with an average concentration of $\approx$0.16. For the slowest t_0, it takes almost 3 weeks to reach the maximum mean concentration in the wall, and the mean peak concentration is less than half of the peak concentration for $t_0 = 0.1$ days.

Varying the porosity of the polymer coating ε_0 has a qualitatively and quantitatively similar effect on the drug release as varying t_0 (Fig. 4). The porosity ε_0 enters

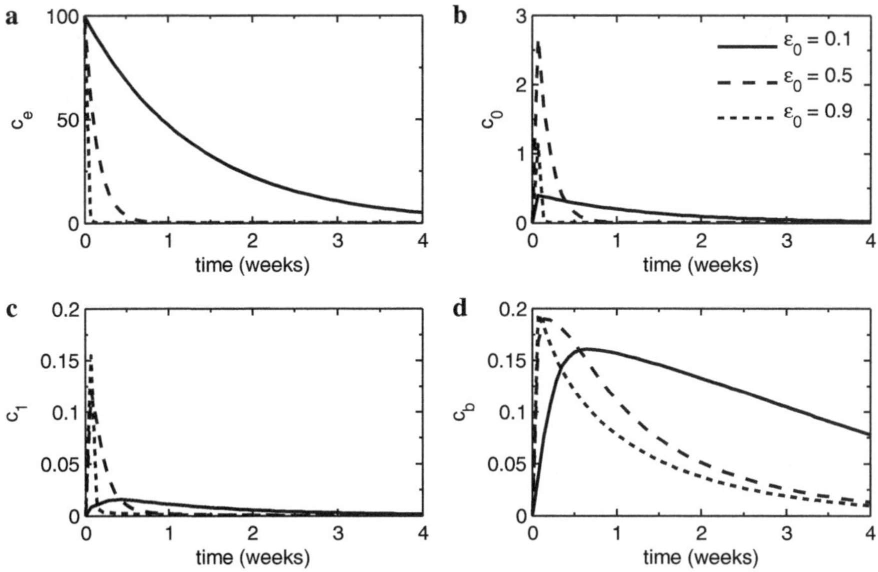

Fig. 4 Sensitivity analysis of the phase-change model in both layers to the porosity of the polymer ε_0. Temporal evolution of **a** the solid-phase drug concentration c_e; **b** the liquid-phase drug concentration c_0; **c** the free drug concentration c_1; **d** the bound drug concentration c_b, each averaged over the entirety of their respective layers

implicitly in our model through the continuity condition between polymer and wall Eq. (10) and in the unbinding rate constant β_0. As the fraction of polymer is reduced and the liquid space is increased, drug release is faster since less drug is stored in the coating (Fig. 4a). However, since we kept the volume-averaged initial concentration constant, the intrinsic drug concentration in the solid phase increases $\left(c_e^s = \dfrac{c_e}{\varepsilon_0}\right)$. This leads to higher liquid phase concentrations in the coating (Fig. 4b). The elevated mean free drug concentration levels in the wall occurring at relatively high porosities indicate that the wall is "overloaded" with free drug causing some drug not to be bound (Fig. 4c). The uptake of drug in the wall is accelerated and the receptors rapidly occupied (Fig. 4d). Due to the sufficiently fast binding kinetics in the wall, virtually all the receptors can be occupied. Nevertheless, since drug supply from the coating reduces quickly, the concentration levels in the wall drop faster.

Figure 5 shows the temporal evolution of the averaged concentrations for varying drug unbinding times t_1. Figure 5a, b indicate that the drug release in the coating is unaffected by the variations in the drug dynamics in the wall (the three curves overlap). Thus, there is no feedback from the wall to the coating. Figure 5c shows that the faster the unbinding time (which also implies a faster binding time, since K_1 is held constant) the higher the overall concentration levels in the wall, as more drug gets bound to the receptors. Also, the mean peak concentration occurs earlier for

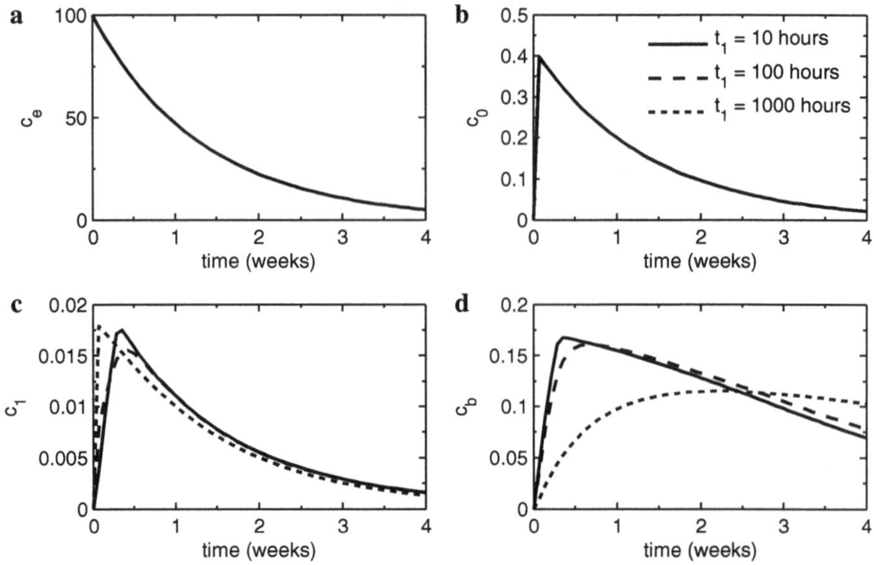

Fig. 5 Sensitivity analysis of the phase-change model in both layers with respect to the characteristic drug unbinding time scale t_1. Temporal evolution of **a** the solid-phase drug concentration c_e; **b** the liquid-phase drug concentration c_0; **c** the free drug concentration c_1; **d** the bound drug concentration c_b, each time averaged over the entirety of their respective layers. In panels **a** and **b** all three curves coincide since the time scale change in the medial layer does not feed back into the polymer layer

faster drug binding/unbinding kinetics. At the same time, the free drug concentration in the wall is barely affected by changes in the drug binding/unbinding kinetics.

On the other hand, if we assume a single liquid-phase c_0 in the coating, we can control the drug release kinetics by setting the drug diffusivity D_0. Using Higuchi's formula, we can estimate the characteristic release time of the drug from the polymer as $t_e = \dfrac{\pi l_0^2}{4D_0}$ [18]. Figure 6a demonstrates how using a single-phase model and varying the characteristic release time t_e (and with that the diffusivity D_0) can in fact be used to approximate the release profiles obtained with the two-phase model varying the solid-liquid transfer time. This leads to very similar bound drug dynamics in the wall (Fig. 6b). These results indicate that a single-phase model in the coating can be used to describe the (averaged) release dynamics and leads to a very similar (averaged) response in the wall. However, from a designer's point of view, the more detailed information about the solid-fluid transfer time and the porosity might be more valuable and easier to control.

Moreover, the current simulations show that, although an additional equation needs to be solved, the proposed two-phase model in the coating is significantly simpler to handle numerically, at least in the one-dimensional case. The reason for this lies in the otherwise very steep concentration gradient at the coating-wall interface. The steep gradient requires a significantly larger number of elements and smaller

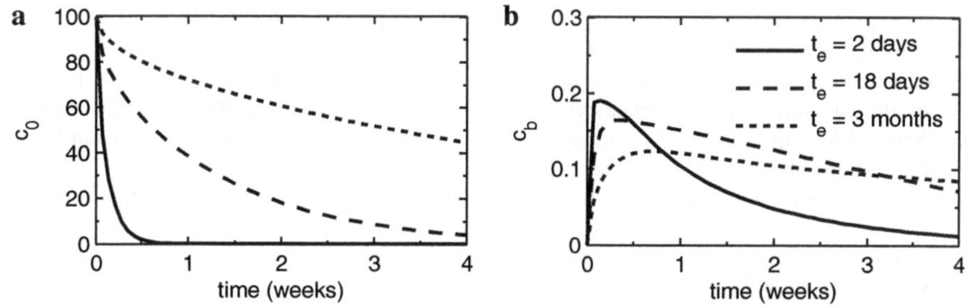

Fig. 6 Sensitivity analysis of the medial layer phase-change model to the characteristic polymer emptying time scale t_e. Temporal evolution of **a** the liquid-phase drug concentration c_0; **b** the bound drug concentration c_b, each time averaged over the entirety of their respective layers

time steps (especially at the beginning of the simulation) which for our numerical method would lead to longer overall calculation times in the case of the single-phase model compared to the two-phase model. The two-phase model avoids these large gradients, since the initial drug is entirely stored in the solid phase at the initial time step, which is not in contact with the wall.

Our previous research has shown that a simplified one-phase model in the wall (assuming instantaneous drug reaction and thus a constant partition of bound and free drug) cannot capture accurately the drug dynamics [15]: neither drug accumulation nor drug residence time can be accurately predicted with a one-phase model.

The present results highlight how the release kinetics in the polymer and the drug dynamics in the wall are highly and nonlinearly coupled. The goal of controlled drug release is to obtain and to maintain sufficiently high drug concentrations in the wall so as to keep their therapeutic effectiveness over an extended period of time. The present work shows that the drug release, controlled by t_0 and ε_0 (other than D_0), needs to be tailored to the specific drug kinetics in the arterial wall in order to reach optimal DES performance.

Acknowledgments Franz Bozsak is supported by a grant from the LaSIPS Laboratory of Excellence. This work is supported in part by a permanent endowment in Cardiovascular Cellular Engineering from the AXA Research Fund (to Abdul I. Barakat) and by the MIUR-CNR project "Interomics", 2014 (to Giuseppe Pontrelli).

References

1. Tzafriri, A.R., Groothuis, A., Price, G.S., Edelman, E.R.: Stent eluting rate determines drug deposition and receptor-mediated effects. J. Controlled Release **161**, 918–926 (2012)
2. Tesfamariam, B.: Drug release kinetics from stent device-based delivery systems. J. Cardiovasc. Pharmacol. **51**(2), 118–125 (2008)
3. Venkatraman, S., Boey, F.: Release profiles in drug-eluting stents: issues and uncertainties. J. Control. Release **120**(3), 149–160 (2007)

4. Hwang, C., Wu, D., Edelman, E.R.: Physiological transport forces govern drug distribution for stent- based delivery. Circulation **104**(5), 600–605 (2001)
5. Zunino, P., D'Angelo, C., et al.: Numerical simulation of drug eluting coronary stents: mechanics, fluid dynamics and drug release. Comp. Methods Appl. Mech. Eng. **198**, 3633–3644 (2009)
6. Weiler, J.M., Sparrow, E.M., Ramazani, R.: Mass transfer by advection and diffusion from a drug-eluting stent. Int. J. Heat Mass Transf. **55**, 1–7 (2012)
7. Vairo, G., Cioffi, M., et al.: Drug release from coronary eluting stents: a multidomain approach. J. Biomech. **43**(8), 1580–1589 (2010)
8. McGinty, S., McKee, S., Wadsworth, R.M., McCormick, C.: Modelling drug-eluting stents. Math. Med. Biol. **28**, 1–29 (2011)
9. Hossainy, S., Prabhu, S.: A mathematical model for predicting drug release from a biodurable drug-eluting stent coating. J. Biomed Mater. Res. A **87**(2), 487–493 (2008)
10. de Monte, F., Pontrelli, G., Becker, S.M.: Drug release in biological tissues, Chapt. 3. In: Becker, S.M., Kuznetsov, A.V. (eds.) Transport in Biological Media, pp. 59–118. Elsevier, New York (2013)
11. Aragon, J., Kar, S., et al.: The effect of variable release kinetics on paclitaxel efficacy from a drug eluting stent in a porcine model. EuroIntervention **1**(2), 228–235 (2005)
12. Guagliumi, G., Ikejima, H., et al.: Impact of drug release kinetics on vascular response to different zotarolimus-eluting stents implanted in patients with long coronary tenoses: The LongOCT study (optical coherence tomography in long lesions). JACC Cardiovasc. Interv. **4**(7), 778–785 (2011)
13. Creel, C.J., Lovich, M.A., Edelman, E.R.: Arterial paclitaxel distribution and deposition. Circ. Res. **86**(8), 879–884 (2000)
14. Tzafriri, A.R., Levin, A.D., Edelman, R.: Diffusion-limited binding explains binary dose response for local arterial and tumour drug delivery. Cell Prolif. **42**(3), 348–363 (2009)
15. Bozsak, F., Chomaz, J.-M., Barakat, A.I.: Modeling the transport of drugs eluted from stents: physical phenomena driving drug distribution in the arterial wall. Biomech. Model. Mechanobiol. **13**(2), 327–347 (2014)
16. Pontrelli, G., de Monte, F.: A multi-layered porous wall model for coronary drug-eluting stents. Int. J. Heat Mass Transf. **53**, 3629–3637 (2010)
17. Pontrelli, G., Di Mascio, A., de Monte, F.: Local mass non-equilibrium dynamics in multilayered porous media: application to the drug-eluting stent. Int. J. Heat Mass Transf. **66**, 844–854 (2013)
18. Vergara, C., Zunino, P.: Multiscale boundary conditions for drug release from cardiovascular stents. Multiscale Model. Simul. **7**(2), 565–588 (2008)

Development of Magnesium Alloy Scaffolds to Support Biological Myocardial Grafts: A Finite Element Investigation

Martin Weidling, Silke Besdo, Tobias Schilling, Michael Bauer, Thomas Hassel, Friedrich-Wilhelm Bach, Hans Jürgen Maier, Jacques Lamon, Axel Haverich and Peter Wriggers

Abstract Lesioned myocardial tissue can be replaced with innovative biological grafts. However, the strength of most biological grafts is initially not sufficient for left ventricular applications. Implants that mechanically support these grafts and gradually lose their function as the graft develops its strength are a possible solution. We are developing magnesium alloy scaffolds for this purpose. The finite element method was used to perform simulations wherein scaffolds are deformed according to the heart movement. This allows us to identify highly stressed regions within

F.-W. Bach—Deceased

M. Weidling (✉) · S. Besdo · P. Wriggers
Institute of Continuum Mechanics, Leibniz Universität Hannover, Hannover, Germany
e-mail: weidling@ikm.uni-hannover.de

S. Besdo
e-mail: s.besdo@ikm.uni-hannover.de

P. Wriggers
e-mail: wriggers@ikm.uni-hannover.de

T. Schilling · A. Haverich
Transplantation and Vascular Surgery, Hannover Medical School, Hannover, Germany
e-mail: schilling.tobias@mh-hannover.de

A. Haverich
e-mail: haverich.axel@mh-hannover.de

M. Bauer · T. Hassel · F.-W. Bach · H.J. Maier
Institut Für Werkstoffkunde (Materials Science), Leibniz Universität Hannover, Hannover, Germany
e-mail: bauer@iw.uni-hannover.de

T. Hassel
e-mail: hassel@iw.uni-hannover.de

H.J. Maier
e-mail: maier@iw.uni-hannover.de

J. Lamon
Laboratoire de Mécanique et Technologie, École Normale Supérieur de Cachan, Cachan, France
e-mail: lamon@lmt.ens-cachan.fr

T. Lenarz and P. Wriggers (eds.), *Biomedical Technology*, Lecture Notes in Applied and Computational Mechanics 74, DOI 10.1007/978-3-319-10981-7_6

the implant that need design changes. Preformed scaffolds were determined to have significantly lower stresses in comparison to flat ones. The method of tensile triangles suggests shape changes for notable stress reduction. Furthermore, new scaffold shapes were developed and simulated. Two of them are recommended for further examinations through in vitro and in vivo tests. A completely new alternative scaffold concept is also proposed.

Keywords Numerical simulation · FEM · Tensile triangles · LA63 · Left ventricle · Heart attack · Tissue substitution · Supporting structure

1 Introduction

Despite the fact that the incidence of severe coronary artery diseases is decreasing in industrial nations, they are still one of the leading causes of death. Approximately 280,000 heart attacks occur every single year in Germany alone. A heart attack does not necessarily lead to death, but it will likely cause long term negative effects on the heart. Stenosed arteries can result in death of the myocardial tissue they supply. This necrotized tissue will be replaced by scar tissue. A loss of myocardial tissue means a loss of contractile tissue, resulting in reduced heart performance, which eventually results in congestive heart failure. Congestive heart failure often leads to a secondary end organ failure such as liver insufficiency, renal failure, and severe edema. Surgical treatment is required to restore sufficient hemodynamic performance of the heart. Therefore, current available techniques strive to restore the physiological shape and volume capacity of the ventricle [1–3]. However, reconstruction of large lesioned areas of the ventricle still results in a less than optimal ejection fraction [4, 5]. Thus, the ideal myocardial reconstructive therapy would utilize regenerative tissue prostheses that could restore full contractile ability of the lesioned cardiac tissue. One research direction is to reduce the scar tissue size by injecting cardiac stem cells and bone marrow mesenchymal stem cells [6]. Even though a slight boost of the heart's performance can be detected, this cell based enhancement most likely will remain a secondary treatment. However, replacement of the scar tissue with solid innovative and regenerative grafts promises to provide a more effective solution that will be available sooner [7].

The use of artificial grafts lacks the problems of non-contractility and susceptibility to infections. Biological grafts appear to be a preferable solution. A few tissues showed a good potential, such as the urinary bladder [8], pericardium [9], skeletal muscle [10] or myocardium [11]. Tudorache et al. had success with autologous small intestine segments without mucosa [12]. After substitution of a limited area of right ventricular tissue with vascularized autologous small intestine, repopulation of these decellularised grafts with cardiomyocytes was shown in animal studies. This gave evidence of myocardial remodeling processes [12]. In a similar approach, Badylak et al. detected cardiomyocytes in urinary bladder matrixes used as a reconstructive myocardial patch [13]. However, these studies were done for use on the right ventricle, with blood pressures of about 40 to 60 mmHg. These grafts are not recommended for use with the left ventricle, which has blood pressures of up to 240 mmHg. Ruptures are likely due to the insufficient strength of those biological grafts, especially in the early stages after implantation. Some additional degradable implants could help to mechanically support the grafts until sufficient strength is obtained through in vivo remodeling processes.

Different degradable materials are used in clinical research. A few ceramics can degrade, for instance the group of calcium phosphates [14]. However, ceramics present the problem of high strength in combination with low ductility. For the intended implants, high ductility is a necessary characteristic. Some polymers degrade into nontoxic substances in the physiological environment, but polymers have low strength and are limited to low stress applications due to their creep behavior [15, 16]. Furthermore, inflammation was observed in soft tissues after implantation of polymeric grafts [17]. Metal implants appear as an alternative. Magnesium and its alloys have been the particular focus of recent research [18]. Their mechanical properties are preferable to those of ceramics and polymers. Various alloys were developed as implant materials [19]; furthermore, it seems possible to customize degradation performance for specific applications [20]. Since magnesium is part of the physiological metabolism, it can be seen as nontoxic in physiological concentrations [21]. Magnesium alloy scaffolds can support biological grafts until they achieve sufficient strength [22]. Still, the biocompatibility of such alloys needs to be tested in order to investigate the influence of alloying elements and corrosion products on the host's tissue [23, 24].

To this end, scaffolds made from magnesium alloy LA63 were implanted epicardially and tested for in vivo behavior in a porcine model. All investigations performed demonstrated high biocompatibility. Nevertheless, fractures in the scaffolds were detected one month after implantation [24]. Flat scaffolds were used, which the surgeon manually deformed to match the heart curvature shape before sewing them onto the myocardium. It is very likely that the material suffered damage due to high strains during the deformation process. It was assumed that preoperative adaptation of the scaffold to the heart curvature would extend its durability. In vitro tests showed a 53 % increase in the lifetime of preformed scaffolds as compared to the flat scaffolds [25]. Presently, in vivo tests of the preformed scaffolds have not been carried out. Based on the assumption that biological grafts develop sufficient strength after about three months, we still see potential to improve our scaffold designs for an extended durability.

The manual development of scaffold designs using a 'trial and error' method—manufacturing specimens and testing them in vitro and in vivo—is very expensive. Instead, the use of standard engineering tools like CAE[1] is recommended to improve the development process and therefore save time and costs.

A finite element model is developed to mimic the heart's left ventricular movement. Using this, a designed scaffold is tested and highly stressed and strained regions are identified. Based on that knowledge, the scaffold shape is improved and new shape

[1] Computer Aided Engineering.

ideas are introduced. Avoiding the occurrence of plastic strain is necessary. Furthermore, stresses within the scaffold should be as low as possible, because stresses can accelerate magnesium corrosion [26]. The intention of this research is to propose an improved scaffold design with the potential for extended durability. This design will need to be verified through in vitro and in vivo tests.

2 Simulation of Flat and Preformed Scaffolds

2.1 Finite Element Modeling

2.1.1 Heart Movement

A heart attack can occur in different regions of the myocardium. In this feasibility study, we focused on the anterior basal area because of its almost planar surface topography. Based on values of magnetic resonance imaging (MRI) of 28 patients, the average heart curvature was measured in two different planes; averaged values are given in Table 1. Formulas of circular segments were used to define the following displacements:

$$\Delta x = r_{sa} \sin\left(\frac{x}{r_{sa}}\right) - x \tag{1}$$

$$\Delta y = r_{va} \sin\left(\frac{y}{r_{va}}\right) - y \tag{2}$$

$$\Delta z = \mathrm{r_{sa}}\left(\cos\left(\frac{\mathrm{x}}{\mathrm{r_{sa}}}\right) - 1\right) + \mathrm{r_{va}}\left(\cos\left(\frac{\mathrm{y}}{\mathrm{r_{va}}}\right) - 1\right). \tag{3}$$

With these, a circular plate can be deformed into a double curved shape as shown in Fig. 1. This represents a simplified model of a myocardium, whereby heart movement is simulated through changes between the radii of end diastolic and end systolic state.

Table 1 Averaged anterior basal heart curvature [25]

Heart radii	Vertical long axis ($\mathrm{r_{va}}$) (mm)	Short axis plane ($\mathrm{r_{sa}}$) (mm)
End diastolic	49.7	25.1
End systolic	45.7	21.3
Mean value (preformed)	47.7	23.2

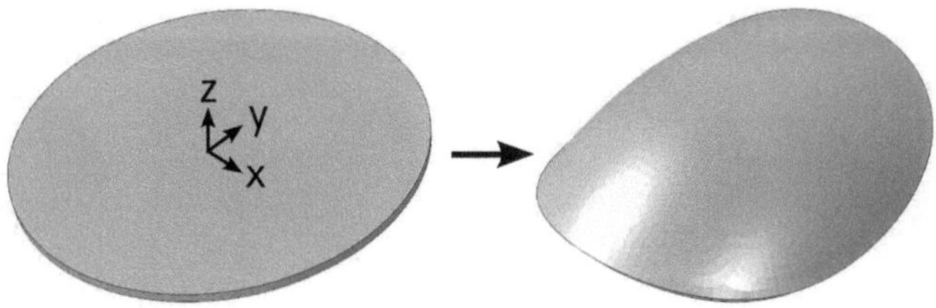

Fig. 1 Modeling of myocardium by deforming a plate according to the heart curvature

2.1.2 Element Type

Simulations were performed with the commercial code Abaqus/CAE (Dassault Systèmes Simulia Corp., RI, USA). Abaqus offers different element types for structural problems. Since the response of the scaffolds are bending dominated special elements had to be selected. Standard linear hexahedral elements appear to be inappropriate; instead the linear formulation with incompatible mode was used (C3D8I) which yields accurate displacement and stress fields. In a test case four elements were used over the thickness to compare the finite element solution to an analytical solution of a clamped beam. Stress was normalized to the analytical solution and calculated to be 0.982 at the fixed support (C3D8I). For quadratic hexahedral elements, normalized stress was determined to be 0.995, and is equal for reduced and full integration formulations (C3D20R and C3D20). Increasing the element quantity about eight times (C3D8I $\rightarrow$ 0.991; C3D20R & C3D20 $\rightarrow$ 0.999) and also about 35 times (C3D8I $\rightarrow$ 0.995; C3D20R & C3D20 $\rightarrow$ 1) lead to minor improvements in accuracy. Evaluating the normalized tip displacement gives similar results. Since linear elements are less costly in calculations and the element C3D8I has a comparable accuracy to quadratic element types, the linear type is preferable. Furthermore, complex structures can be discretized with a finer mesh without extending the calculation time using linear elements as compared with quadratic elements.

2.1.3 Assembly

Two scenarios were considered: a surgeon manually deforming a flat scaffold according to the heart curvature and an already preformed scaffold sewn onto the myocardium. These modeling ideas were briefly introduced by Bauer et al. but are presented here in more detail [25].

The scaffold is positioned onto the myocardium and single nodes are constrained to avoid its translation and rotation, as shown in Fig. 2a. It is assumed that surgeons use their thumbs to deform the scaffold; therefore, thumb models are positioned onto the scaffold (Fig. 2b). 'Thumbs' are modeled to be equidistant and perpendicular to the myocardium. The scaffold is deformed in two steps as displayed in Fig. 2c, d.

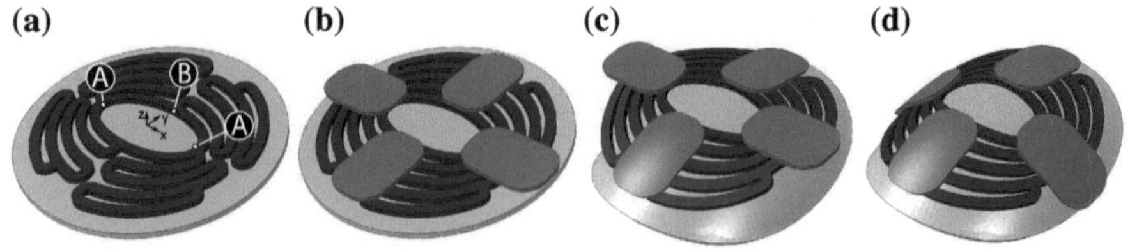

Fig. 2 Modeling of flat scaffold: **a** Positioning of flat scaffold on plate, nodes A are constrained in y-direction and node B is constraint in x-direction; **b** positioning of thumbs onto scaffold; **c** and **d** deforming flat scaffold into heart curvature in two steps

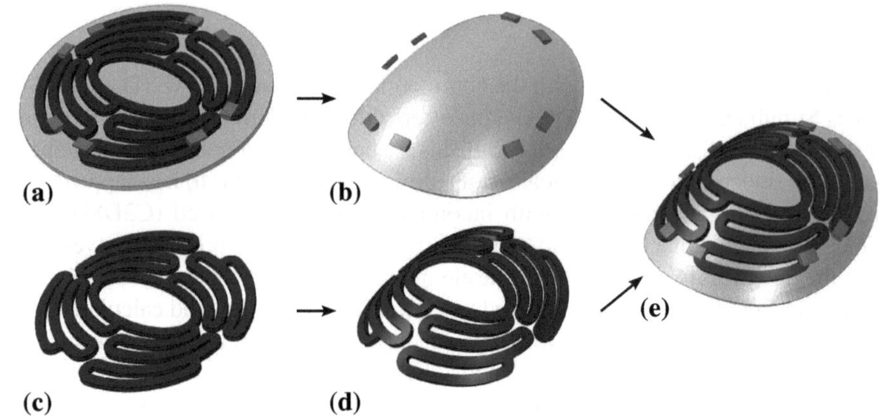

Fig. 3 Modeling of preformed scaffold: **a** positioning of sutures onto flat scaffold; **b** preforming of myocardium with sutures; **c** and **d** deforming flat scaffold into preformed state; **e** assembling of orphan-meshes of scaffold, myocardium and sutures in preformed state

Preformed scaffolds can be designed with CAD[2]-programs and imported into finite element (FE) software. However, it is costly to mesh curved, complex geometries with hexahedral elements; furthermore, positioning of sutures is challenging. Therefore, another method was chosen. In flat state, sutures were positioned on the scaffold according to their locations in surgeries (Fig. 3a). Sutures were simplified as half cylinders. They restrain their position relative to the myocardium during heart movement. The myocardium with sutures was deformed into the preformed state (see Table 1) as shown in Fig. 3b. In the second step, the scaffold was also deformed into the preformed shape (Fig. 3c, d). The preformed meshes are imported as orphan-mesh into a second model (Fig. 3e). Orphan-meshes contain no history data; therefore, they are stress free, comparable to a preformed structure after heat treatment. The heart movement of the preformed myocardium is modeled using the displacements in Eqs. (1–3), but it is necessary to calculate the displacement differences from one curvature state to the other.

[2] Computer Aided Design.

Table 2 True stress and true plastic strain data of Mg-alloy LA63 [25]

True stress (MPa)	137	150	174	187	197
True plastic strain	0.0000	0.0020	0.0107	0.0182	0.0263

2.1.4 Parameters

The magnesium alloy LA63 is not commercially available. Its properties were determined through tensile testing, and found to be: Young's modulus of 46,068 MPa and Poission's ratio of 0.28. Nonlinear material parameters are given in Table 2; isotropic hardening was assumed. Myocardium and sutures as well as thumbs were assumed to be linearly elastic and isotropic, Young's moduli were chosen as 66 MPa [27] and 1,300 MPa, for the myocardium and sutures (thumbs), respectively.

Surface-to-surface contact was defined between interacting bodies, and frictionless contact was assumed. The penalty method was used as contact algorithm with a stiffness scale factor of 0.01.

Components were meshed with linear hexahedral elements with incompatible modes (C3D8I). A sensitivity analysis was performed to choose an appropriate mesh size. Accordingly, the model of the preformed scaffold was meshed with different refinements. Von Mises stresses were determined at five points and compared to the element sizes as shown in Fig. 4. Further, the computing time was contrasted. Stress changes were observed at points 1 and 2, while all other locations were unaffected. These two points are in or close to tight radii where higher stress concentrations can be assumed; therefore a higher sensitivity to the mesh size is observed. An element size of '0.2 mm × 6' was chosen in order to adequately account for sharp corners and complex geometries while still maintaining a reasonable computing time. Therefore, a 1 mm thick scaffold was meshed with an element edge size of 0.2 mm and six elements in thickness.

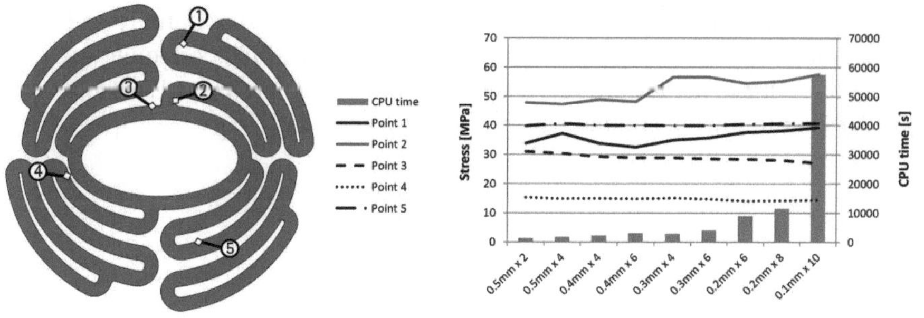

Fig. 4 Sensitivity study on mesh refinements and resultant stress in five points

2.1.5 Simulations

In the first simulation the response of a flat 1 mm thick scaffold was investigated that is deformed by a surgeon to match the shape of the heart curvature before it is sewn onto the myocardium. For this the 'thumb'-model from Fig. 2 was used and the myocardium was deformed into the preformed state. The second simulation examined the behavior of a preoperative preformed scaffold (stress free) that is deformed from preformed to end diastolic to end systolic state. This simulation of one heart movement cycle was performed for scaffolds of 0.5 mm and 1 mm thickness.

2.2 Modeling Results

The flat scaffold was deformed in two steps by thumbs and then kept in a position equivalent to the preformed state. Relatively high stresses occurred in the scaffold, as can be seen in Fig. 5. The yield point of 137 MPa was exceeded in several areas. Furthermore, stresses above the ultimate tensile strength of 197 MPa were determined for a few locations. Stresses beyond the yield point indicate plastic deformations, which are displayed in Fig. 5. The plastic strain that corresponds to the ultimate tensile strength was determined to be 2.6 %. This value was used as scale limit; therefore, strains exceeding it are plotted in red in Fig. 5. Areas stressed and strained above the point of ultimate tensile strength are susceptible to material overload, therefore material failure is likely. Further, plastic strains during cyclic loading can lead to fatigue.

The stress distribution of preformed scaffolds, deformed into end diastolic and into end systolic state, is shown in Fig. 6. Compared to the flat scaffold, calculated stresses were much lower. Maximal stress in the 0.5 mm thick scaffold was determined to be 26 MPa and 59 MPa for the end diastolic and end systolic states, respectively. For the 1 mm thick scaffold maximal stress was higher: 45 MPa in end diastolic and 96 MPa in end systolic state. No plastic strains were observed.

From these results, it can be concluded that it is advisable to use initially preformed scaffolds instead of flat ones. Material stresses in flat scaffolds are quite high and early failures due to fatigue are very likely. Therefore, only preformed scaffolds were considered in following simulations.

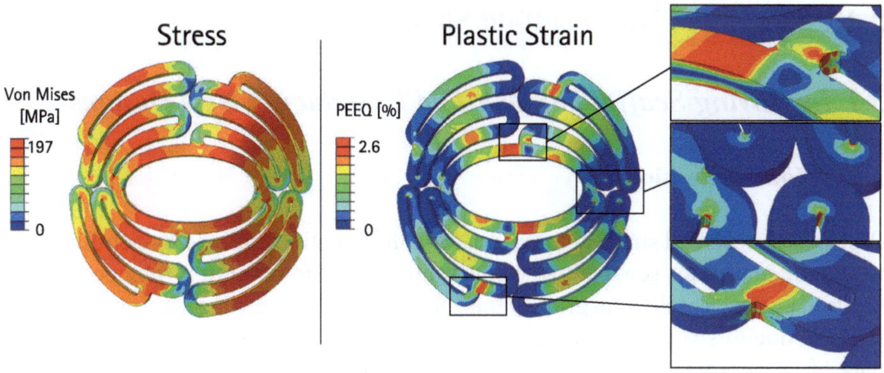

Fig. 5 Distribution of stresses and plastic strains within a 1 mm thick flat scaffold deformed into preformed state

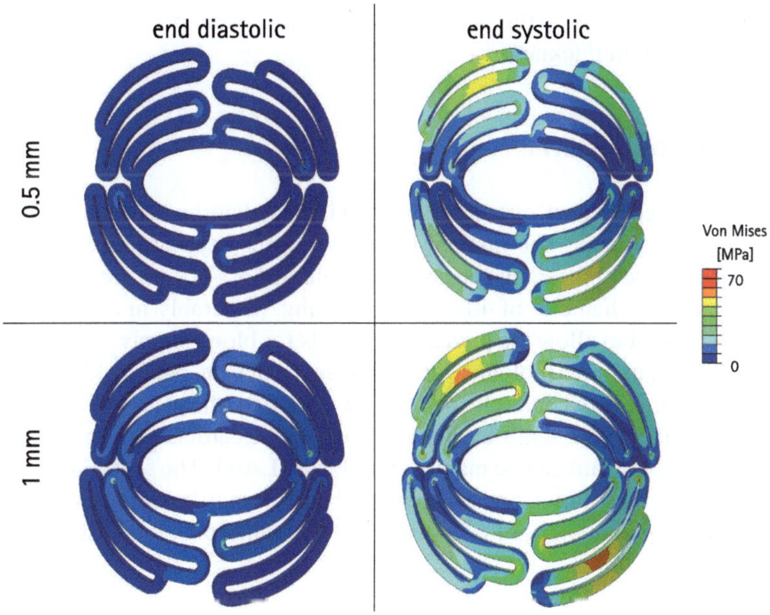

Fig. 6 Stress distribution in end diastolic and in end systolic state for 0.5 mm and 1 mm thick preformed scaffolds

3 Developing of New Scaffold Shapes

3.1 Improving Scaffold Shapes and Introducing New Designs

3.1.1 Model Extension

Previously presented simulations correlate to the situation outlined by Schilling et al. where the scaffold is sewn onto the epicardium [24]. However, it is intended that the scaffold protect a biological graft against bulging. Therefore, the resulting load on the scaffold due to swelling of the graft must be taken into account. Since no material data of an appropriate tissue was available, another approach was chosen. It was assumed that a bulging graft induced a pressure onto the scaffold. Pressure peaks of up to 240 mmHg are possible in the left ventricle. For the simulation, 140 mmHg was chosen—representing prehypertension— and applied onto the back of the scaffold in the systolic step [28]. With this approach, calculations were performed for scaffolds of 0.5 mm and 1 mm thickness.

3.1.2 Design Requirements

Manufacturing techniques, operating site, and intended purpose of an implant define its design requirements. The implant is for use on the myocardium, while its size depends on the size of the damaged area. Diameters of 30–60 mm seem to be typical; we chose an outer diameter of 40 mm. An opening, preferably in the middle of the scaffold, is needed to allow the passage of vessels for blood supply. The intention of the scaffold is to support a biological graft and prevent it from bulging [22]. Therefore, a high contact area between graft and implant is needed, the heart movement should not be restricted, and the scaffold should withstand the cardiac motion. Sheets down to a thickness of 0.5 mm can be manufactured from LA63. The scaffold should have strut widths of at least 1 mm. Spaces between the struts may be a minimum of 0.6 mm due to the cutting diameter of the abrasive waterjet cutting method used to form the scaffolds [29].

3.1.3 Shape Improvements

Under pure bending, a reduction of structure thickness will lead to lower maximal stress. This effect was already investigated (see Fig. 6). Further, it is possible to locally decrease the thickness where high bending stresses are detected and thereby introduce a kind of 'hinge'. However, this approach increases the manufacturing expense and is not advocated. A uniform structure thickness was chosen, with stresses reduced by changing the structure shape.

Mattheck suggests the use of 'tensile triangles' [30]. Using this method, a corner is shaped as shown in the construction in Fig. 7. Starting with the left triangle each

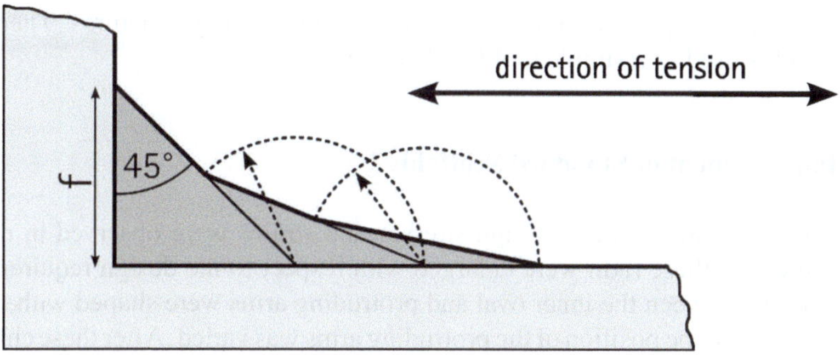

Fig. 7 Construction of tensile triangles according to Mattheck [30]

Table 3 Normalized stress in rounding compared to tensile triangle

Radius 0.5 (mm)	Radius 1 (mm)	Chamfer 0.5 (mm)	Chamfer 1 (mm)
1.21	1.11	1.17	1.32

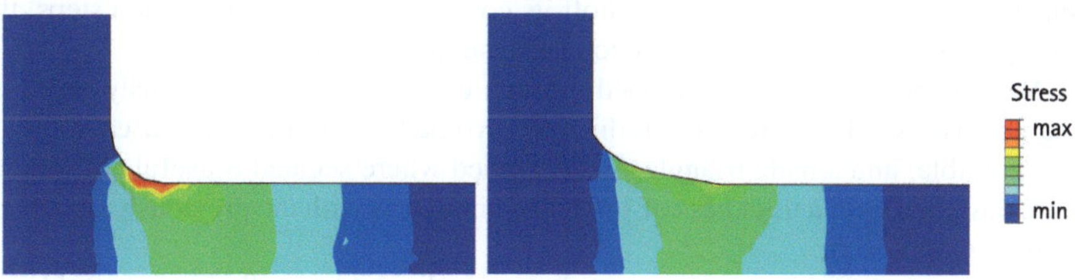

Fig. 8 Stress distribution in a rounding with radius of 0.5 mm (*left*) and a rounding shaped with tensile triangles with a basic length of 0.5 mm (*right*)

subsequent triangle begins in the middle of the previous one and extends in the direction of tension. In most cases, three triangles are sufficient. Obtuse corners are rounded off with radii, except the corner of the first triangle.

This method is described in [30] for shape improvements in the direction of tension; here we applied and tested it for bending dominated problems. Therefore, a 1 mm thick cantilever beam with elastic material properties was modeled. The thickness is unchanged, but corners in width are modified. One 90° corner was shaped with tensile triangles, basic length of $f = 0.5$ mm, and compared with corners formed with radii of 0.5 mm and 1 mm, as well as corners with chamfers of 0.5 mm and 1 mm. The beam tip was deflected. Maximal von Mises stress was determined and normalized to the tensile triangles model as given in Table 3. Further, the stress distribution of the tensile triangles is displayed in comparison with the 0.5 mm radius in Fig. 8.

With tensile triangles, the maximal stress is reduced about 21 % in comparison with a radius of 0.5 mm as stated in Table 3. Also, a much smoother stress distribution is obtained, as can be seen in Fig. 8. These results are promising: the method of tensile triangles is appropriate to reduce stresses in bending structures. Therefore, this method is used as a simple tool for shape improvements.

3.1.4 Improvement of Standard Scaffolds

For previously tested scaffolds, high stresses and strains were observed in narrow radii. Therefore, these radii were enlarged with respect to the design requirements. The junctions between the inner oval and protruding arms were shaped with tensile triangles. Further, the position of the protruding arms was varied. After these changes, the previously presented scaffold was referred to as the 'standard scaffold'.

3.1.5 New Designs

The method of tensile triangles gives an idea of an optimized shape, but makes no predictions about its sizing. A scaffold in the shape of a snowflake was designed and tensile triangles were used to round off its corners. In several calculation steps, the tensile triangles were adjusted to decrease resultant stresses.

New shape ideas were developed through consideration of previously explored characteristics. Therefore, tight radii were avoided, large arcs were used where it was possible, and tensile triangles were applied where seemed it useful. All shapes were designed with a thickness of 1 mm and simulated with the previously described modeling approach.

One-piece scaffolds were presented previously. Additionally, one scaffold was designed consisting of several segments as shown in Fig. 9. The segments can be linked by degradable polymers, e.g. resorbable sutures.

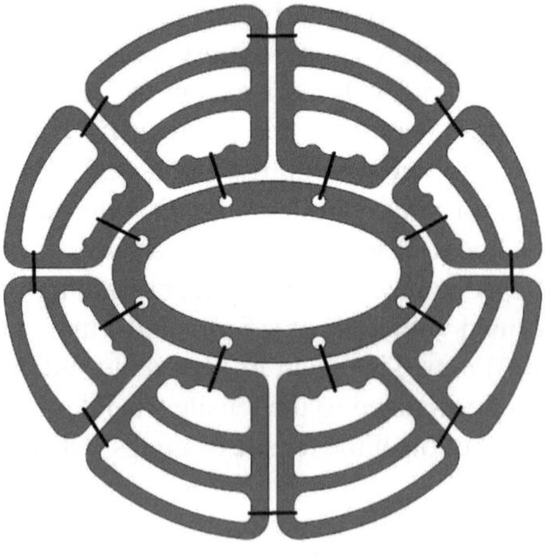

Fig. 9 Myocardial scaffold consisting of segments that are linked by degradable sutures

3.2 Results for Shape Improvements and New Designs

Extending the simulation and adding pressure onto the back of the scaffold in systolic step had an impact on results. Stresses in the diastolic step were still the same as presented in Fig. 6, but stresses in the systolic step increased significantly, as can be seen in Fig. 10. The 1 mm thick scaffold had a maximum stress of 164 MPa, while the maximal von Mises stress of the 0.5 mm implant was determined to be 199 MPa. For both scaffolds, plastic strains occurred in end systolic state.

Including a systolic pressure had a more profound influence on the thinner scaffold; its lower stiffness caused greater deformations. This can be seen in Fig. 11, where the 0.5 mm scaffold is lifted from the myocardium. The 1 mm thick scaffold is less affected by the systolic pressure and had only minor lifting. Nevertheless, its maximal stress with 164 MPa was about 70 % higher in comparison with the model without a systolic pressure. Further, the stress distribution changed, as can be seen in Figs. 6 and 10. Consequently, modeling of a systolic pressure was included in all following calculations to cover bulging grafts.

With the broadening of narrow radii, a slight decrease in maximal stress was reached but the highest stresses were still achieved at the inner tight radii, as pictured in Fig. 12. In contrast to the standard scaffold, no plastic strains were observed. Keeping the design requirements in mind, it seems impossible to adjust the standard

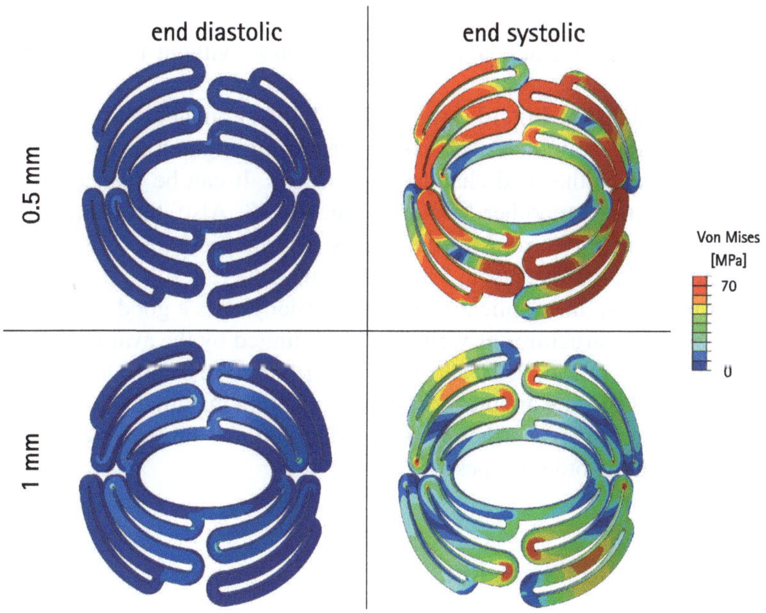

Fig. 10 Stress distribution in end diastolic and in end systolic state for 0.5 and 1 mm thick standard scaffolds after including a systolic pressure

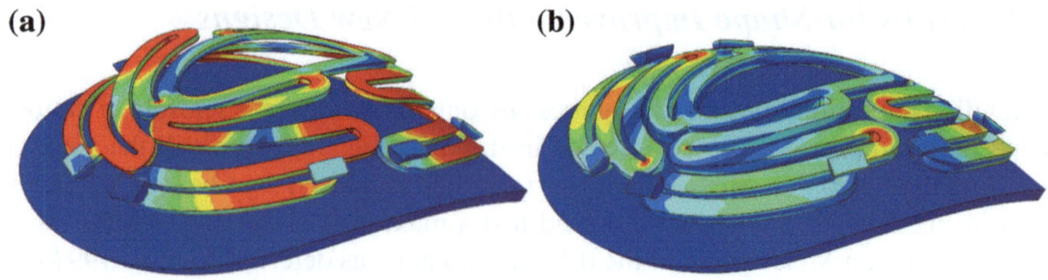

Fig. 11 View of myocardium with standard scaffold in end systolic state: **a** 0.5 mm thick scaffold; **b** 1 mm thick scaffold

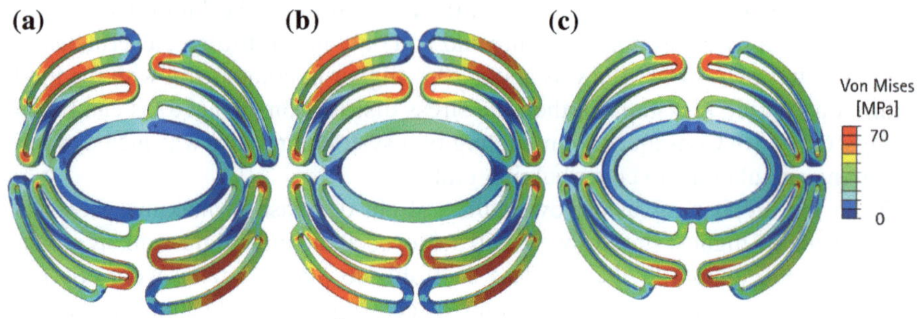

Fig. 12 Shape improvements and variants of standard scaffold

scaffold shape to achieve adequate stress distribution without fundamental design changes.

The scaffold shape in Fig. 13a was designed and simulated with the previously described modeling approach. The corner of the highlighted arm was structured with tensile triangles that had changed dimensions. It can be seen that the stress distribution is getting smoother from b to e in Fig. 13. Also, the maximal stress is decreased by about 11 % from b to e. This was achieved by increasing the basic length of the tensile triangles from 5 mm to 9 mm, and as a result the entire rounding grew. For this reason, the application of tensile triangles is a good option to reduce stresses in complex structures as well. It is only limited by the available space.

Only the shapes 1, 2, 6, 9, 10 and 12 in Fig. 14 are free of plastic strains. Out of these, shape 6 and shape 9 are the most promising, with maximal stresses of 90 MPa and 120 MPa, respectively. Further, the stress distribution is smoother in these designs as compared with the other shapes.

4 Summary and Discussion

Loss of myocardial tissue due to ischemic incidents results in reduced heart performance. Innovative biological grafts can be used to replace lesioned tissues [7]. Furthermore, their recolonization with cardiomyocytes gives evidence supporting

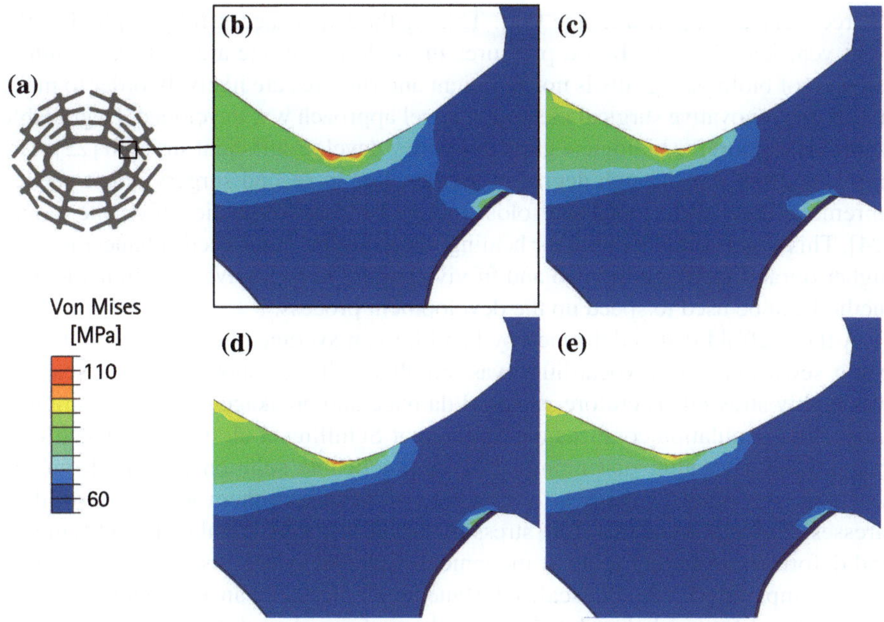

Fig. 13 **a** Stress distribution in a rounding after increasing of tensile triangles basic length f from **b** 5 mm, to **c** 7 mm, to **d** 8 mm, to **e** 9 mm

Fig. 14 Stress distribution in newly developed scaffold shapes

the recovery heart performance [12]. This method was successfully applied to the right ventricle [12, 13]. Blood pressures in the left ventricle are higher; the initial strength of biological grafts is not sufficient and ruptures are likely. In order to make use of this innovative surgical method, a novel approach was developed. Degradable scaffolds support the biological grafts until they develop sufficient strength [22]. The implants disappear through degradation; therefore, a second surgery is not needed to remove them. Developed scaffolds fractured in in vivo studies after one month [24]. This is not sufficient time for healing; therefore implants need enhancement for higher durability. Since in vitro and in vivo testing is expensive, the finite element method can be used to speed up the development process.

A flat scaffold that is deformed by hand into an averaged heart curvature before being sewn onto the myocardium was simulated. It was shown that the scaffold was highly stressed. Therefore, material damage and breakage of the structure was likely. The simulation confirms the findings of Schilling et al., where flat scaffolds fractured in vivo after one month [24]. A practicable solution is to preform flat scaffolds according to the heart curvature and heat-treat them to reduce residual stresses. Preformed and therefore stress free scaffolds were simulated in a FE model and deformed according to heart movement. Calculated stresses were significantly lower compared with the flat scaffold. Bauer et al. measured an approximately 53 % extension in life-time before breakage for the preformed implant, confirming the FE results [25].

Complex FE models of the left ventricle are described in the literature [31, 32]. Some groups include patient-specific fiber orientation maps in their heart models [33], or simulate filling dynamics of the left ventricle [34]. These models allow investigation of cardiac biomechanics and help in understanding heart diseases [32, 34]. However, for the development of myocardial scaffolds, those kinds of models are too complex. The simulation was limited to modeling the contraction and relaxation of the heart. To accomplish this, the anterior basal heart curvature was measured in 28 patients and average values were calculated [25]. Using these averages, myocardial movement can be mimicked in end diastolic and end systolic states. Only the translational motion of the heart is considered; the rotational motion is neglected. This can lead to an underestimation of the implant loading.

Loads that result from a bulging graft were simulated as pressure on the back of the scaffold in the systolic step. The 0.5 mm thick scaffolds showed that the effect of a bulging graft needs to be taken into account. Simulations without this systolic pressure led to low stresses, because the scaffold was only deformed due to heart curvature changes. After inclusion of the systolic pressure, the scaffold was additionally loaded. Thus, high stresses and larger deformations followed. Physiological pressures of up to 240 mmHg are possible for the left ventricle. To avoid an overestimation of that effect, an averaged pressure of 140 mmHg was chosen for the simulation. In order to enhance the finite element modeling, biological grafts must be included in order to simulate bulging of the graft due to the high blood pressure and thus the deformation of the scaffold.

Complex heart movement and the effect of a sewn on graft were simplified to a translational curvature change with systolic pressure. Therefore, it is not possible to exactly predict the stress and strain situation of the loaded scaffolds, but tendencies can be observed. A consistent modeling technique was introduced, allowing comparisons and assessment of different scaffold shapes.

It was shown that flat scaffolds fail too early and are therefore not regarded anymore. Preforming of the scaffolds seems to be a promising improvement concept and was applied to all designs. According to the used modeling technique, scaffolds of 0.5 mm thickness appear to be inappropriate because they are not stiff enough to constrain a bulging graft. Design changes to the standard scaffold brought minor improvements, but extensive design changes were needed. The method of tensile triangles was introduced and tested [30]. Tensile triangles are a promising tool and were applied to scaffolds where it seemed appropriate. Complex structures sometimes needed several iterations of shape adjustments. Also, the quantity of stress reduction depends on the available space to enlarge tensile triangles. Furthermore, new scaffold designs were introduced. A few of them appear promising. For shape 6 and shape 9, maximal von Mises stress was reduced by about 45 and 27 %, respectively, in comparison with the standard scaffold. The benefit of the improved scaffolds needs to be tested in vitro.

Plastic strains in a structure under cyclic loading can lead to very fast fatigue. Therefore, our goal was to design scaffolds that are strained in the elastic range. However, stresses in elastic range must still be monitored. Stresses can accelerate corrosion [26]. For this reason, highly stressed areas of a specimen are attacked faster. Further, stress corrosion cracking (SCC) can lead to rapid fracture [35]. Many magnesium alloys have a threshold stress for SCC on the order of 50 % of the yield stress [36]. The yield point of LA63 is at 137 MPa, but all developed scaffolds were loaded above 70 MPa. Furthermore, fatigue is also related to stresses in the elastic range. Presently, no endurance tests have been performed on the in-house alloy LA63. However, for several aluminum containing magnesium alloys, the endurance limit was below 70 MPa for 10^7 cycles [37, 38]. Consequently, the superposition of both effects—fatigue and corrosion—can lead to a premature material failure.

Future work will include enhancing the simulation by including a bulging graft. With this, the most promising scaffolds will be assessed again. Results will be compared with in vitro tests. If the stresses are still too high, then the concept of a one-piece scaffold must be reconsidered. An alternative scaffold consisting of segments was already presented. This implant is expected to have significantly lower stresses. Therefore, the risk of fatigue and also of accelerated corrosion due to high stresses is decreased. One disadvantage of this scaffold type is the laborious handling in surgery required for the movable segments.

5 Conclusion

A finite element model was developed to simulate myocardial scaffolds that are deformed through cardiac motion. Incidents that were already observed in vitro and in vivo were predictably replicated in this simulation. Hence, flat scaffolds should be avoided; instead preformed structures are recommended. Several new shapes were developed and simulated. For a few of them, significant stress reductions were observed as compared with the initial scaffold. The improvement of the most promising is to be confirmed through in vitro tests. With help of finite element simulations, stresses and strains within scaffolds are predictable; therefore, different shapes can be compared. With this, it is possible to narrow prospective options before costly test series are performed. Consequently, developing time and expenses are reduced.

We also discussed the idea that newly developed scaffolds could possibly be stressed too much due to heart movement. Therefore, it is recommended to enhance the simulation and to validate it with in vitro tests. If stresses within the scaffolds are still too high, another scaffold concept is recommended, see Fig. 9. With this linked scaffold, the handling effort is increased during surgery, but this is acceptable for an implant that fulfills its function and exhibits low fatigue.

Acknowledgments The authors are thankful to the German Research Foundation (DFG) for their financial support. This project is funded within the Collaborative Research Center 599 (SFB 599) and the International Research Training Group 1627 (GRK 1627). Furthermore, we thank Martina Baldrich who developed scaffold shape 7 and Julian Schrader who developed shapes 8–12 in student projects, respectively.

References

1. Athanasuleas, C.L., Stanley, A.W., Buckberg, G.D., et al.: Surgical anterior ventricular endocardial restoration (SAVER) in the dilated remodeled ventricle after anterior myocardial infarction. RESTORE group. Reconstructive Endoventricular Surgery, returning Torsion Original Radius Elliptical Shape to the LV. J. Am. Coll. Cardiol. **37**(5), 1199–1209 (2001)
2. Cooley, D.A.: Ventricular endoaneurysmorrhaphy: results of an improved method of repair. Tex. Heart Inst. J. **16**(2), 72–75 (1989)
3. Dor, V.: Surgery for left ventricular aneurysm. Curr. Opin. Cardiol. **5**(6), 773–780 (1990)
4. Adhyapak, S.M., Parachuri, V.R.: Architecture of the left ventricle: insights for optimal surgical ventricular restoration. Heart Fail. Rev. **15**(1), 73–83 (2010)
5. Anderson, R.H., Ho, S.Y., Redmann, K., et al.: The anatomical arrangement of the myocardial cells making up the ventricular mass. Eur. J. Cardiothorac. Surg. **28**(4), 517–525 (2005)
6. Williams, A.R., Hatzistergos, K.E., Addicott, B., et al.: Enhanced effect of combining human cardiac stem cells and bone marrow mesenchymal stem cells to reduce infarct size and to restore cardiac function after myocardial infarction. Circulation **127**(2), 213–223 (2013)
7. Schilling, T., Cebotari, S., Tudorache, I., et al.: Tissue engineering of vascularized myocardial prosthetic tissue. Chirurg **82**(4), 319–324 (2011). in German
8. Badylak, S.F.: The extracellular matrix as a scaffold for tissue reconstruction. Semin. Cell Dev. Biol. **13**(5), 377–383 (2002)
9. Wei, H., Liang, H., Lee, M., et al.: Construction of varying porous structures in acellular bovine pericardia as a tissue-engineering extracellular matrix. Biomaterials **26**(14), 1905–1913 (2005)
10. Taheri, S.A., Ashraf, H., Merhige, M., et al.: Myoangiogenesis after cell patch cardiomyoplasty and omentopexy in a patient with ischemic cardiomyopathy. Tex Heart Inst. J. **32**(4), 598–601 (2005)
11. Wang, B., Borazjani, A., Tahai, M., et al.: Fabrication of cardiac patch with decellularized porcine myocardial scaffold and bone marrow mononuclear cells. J. Biomed. Mater. Res. **94**(4), 1100–1110 (2010)

12. Tudorache, I., Kostin, S., Meyer, T., et al.: Viable vascularized autologous patch for transmural myocardial reconstruction. Eur. J. Cardiothorac. Surg. **36**(2), 306–311 (2009)
13. Badylak, S.F., Kochupura, P.V., Cohen, I.S., et al.: The use of extracellular matrix as an inductive scaffold for the partial replacement of functional myocardium. Cell Transplant. **15**(Suppl 1), 29–40 (2006)
14. Wiltfang, J., Merten, H.A., Schlegel, K.A., et al.: Degradation characteristics of α and β tricalcium-phosphate (TCP) in minipigs. J. Biomed. Mater. Res. **63**(2), 115–121 (2002)
15. Nair, L.S., Laurencin, C.T.: Biodegradable polymers as biomaterials. Prog. Polym. Sci. **32**(8–9), 762–798 (2007)
16. Pietrzak, W.S., Sarver, D., Verstynen, M.: Bioresorbable implants – practical considerations. Bone **19**(1), 109–119 (1996)
17. van der Giessen, W.J., Lincoff, A.M., Schwartz, R.S., et al.: Marked inflammatory sequelae to implantation of biodegradable and nonbiodegradable polymers in porcine coronary arteries. Circulation **94**(7), 1690–1697 (1996)
18. Atrens, A., Song, G., Cao, F., et al.: Advances in Mg corrosion and research suggestions. J. Magnes. Alloy. **1**(3), 177–200 (2013)
19. Witte, F., Hort, N., Vogt, C., et al.: Degradable biomaterials based on magnesium corrosion. Curr. Opin. Solid State Mater. Sci. **12**(5–6), 63–72 (2008)
20. Kirkland, N., Lespagnol, J., Birbilis, N., et al.: A survey of bio-corrosion rates of magnesium alloys. Corros. Sci. **52**(2), 287–291 (2010)
21. Staiger, M.P., Pietak, A.M., Huadmai, J., et al.: Magnesium and its alloys as orthopedic biomaterials: a review. Biomaterials **27**(9), 1728–1734 (2006)
22. Bach, F., Haverich, A., Cebotari. S., Biskup, C., Schuster, B.: Supporting element for tissue implants (Patent WO 2011/101142 A1)
23. Walker, J., Shadanbaz, S., Woodfield, T.B.F., et al.: The in vitro and in vivo evaluation of the biocompatibility of Mg alloys. Biomed. Mater. **9**(1), 15006 (2014)
24. Schilling, T., Brandes, G., Tudorache, I., et al.: In vivo degradation of magnesium alloy LA63 scaffolds for temporary stabilization of biological myocardial grafts in a swine model. BiomedizinischeTechnik/Biomedical Engineering **58**(5), 407–416 (2013)
25. Bauer, M., Schilling, T., Weidling, M., et al.: Geometric adaption of biodegradable magnesium alloy scaffolds to stabilise biological myocardial grafts, Part I. J. Mater. Sci. Mater. Med. **25**(3), 909–916 (2014)
26. Bonora, P., Andrei, M., Eliezer, A., et al.: Corrosion behaviour of stressed magnesium alloys. Corros. Sci. **44**(4), 729–749 (2002)
27. Hoffmeister, B.K., Handley, S.M., Wickline, S.A., et al.: Ultrasonic determination of the anisotropy of Young's modulus of fixed tendon and fixed myocardium. J. Acoust. Soc. Am. **100**(6), 3933–3940 (1996)
28. Weidling, M., Besdo, S., Schilling, T., et al.: Finite element simulation of myocardial stabilising structures and development of new designs. Biomedical Engineering/BiomedizinischeTechnik. 58 (Suppl. 1)(2013)
29. Biskup, C., Hepke, M., Grittner, N., et al.:AWIJ cutting of structures made of magnesium alloys for the cardiovascular surgery. In: American WJTA Conference and Expo (2009)
30. Mattheck, C.: Secret Design Rules of Nature. Überw. Ill, 1st edn. Forschungszentrum Karlsruhe, Karlsruhe (2007)
31. Feng, B., Veress, A., Sitek, A., et al.: Estimation of mechanical properties from gated SPECT and cine MRI data using a finite-element mechanical model of the left ventricle. IEEE Trans. Nucl. Sci. **48**(3), 725–733 (2001)
32. Feng, L., Weixue, L., Ling, X., et al.: The construction of three-dimensional composite finite element mechanical model of human left ventricle. JSME Int. J. Ser C **44**(1), 125–133 (2001)
33. Wong, J., Kuhl, E.: Generating fibre orientation maps in human heart models using Poisson interpolation. Comput. Methods Biomech. Biomed. Eng.: 1–10 (2012)
34. Watanabe, H., Sugiura, S., Kafuku, H., et al.: Multiphysics simulation of left ventricular filling dynamics using fluid-structure interaction finite element method. Biophys. J. **87**(3), 2074–2085 (2004)

35. Song, G., Atrens, A.: Recent insights into the mechanism of magnesium corrosion and research suggestions. Adv. Eng. Mater. **9**(3), 177–183 (2007)
36. Winzer, N., Atrens, A., Song, G., et al.: A critical review of the stress corrosion cracking (SCC) of Magnesium alloys. Adv. Eng. Mater. **7**(8), 659–693 (2005)
37. Tokaji, K., Kamakura, M., Ishiizumi, Y., et al.: Fatigue behaviour and fracture mechanism of a rolled AZ31 Magnesium alloy. Int. J. Fatigue **26**(11), 1217–1224 (2004)
38. Mayer, H., Papakyriacou, M., Zettl, B., et al.: Influence of porosity on the fatigue limit of die cast magnesium and aluminium alloys. Int. J. Fatigue **25**(3), 245–256 (2003)

Finite Element Analysis of Transcatheter Aortic Valve Implantation in the Presence of Aortic Leaflet Calcifications

Annalisa Dimasi, Marco Stevanella, Emiliano Votta, Francesco Sturla, Gaetano Burriesci and Alberto Redaelli

Abstract Transcatheter Aortic Valve (TAV) implantation is a recent interventional procedure for the replacement of the aortic valve in patients affected by severe aortic stenosis who are considered at high or prohibitive surgical risk. Despite recent improvements, TAV-related complications still limit its application. In the present work, FE analyses of TAV implantation and function have been performed with the aim of investigating the influence of the calcifications of the aortic valve leaflets on TAV performances. Results suggest that the degree and location of calcifications could influence post-implanted TAV configuration as well as TAV-aortic root interactions and TAV dynamics. The study gives insights in the biomechanics of TAV, while the implemented computational tools could be applied to different scenarios to investigate other relevant clinical aspects.

Keywords Transcatheter aortic valve implantation · Finite element analysis · Calcific aortic stenosis

1 Introduction

Transcatheter Aortic Valve (TAV) implantation has been proposed in 2002 as an alternative approach to surgical aortic valve replacement (SAVR) for the substitution of the stenotic aortic valve (AV) in those patients who were denied surgery because of high surgical risk [1]. TAV procedure consists in the percutaneous implantation of a stented biological valve that is delivered in the aortic position through a catheter; the latter is inserted in a periferal vessel in the majority of cases [2]. As mentioned, the prosthetic valve is sewned onto a metallic stent which interacts with the surrounding

A. Dimasi (✉) · M. Stevanella · E. Votta · F. Sturla · A. Redaelli
Department of Electronics, Information and Bioengineering, Politecnico di Milano, Milano, Italy
e-mail: annalisa.dimasi@polimi.it

F. Sturla
Division of Cardiovascular Surgery, Università degli Studi di Verona, Verona, Italy

G. Burriesci
Department of Mechanical Engineering, University College of London, London, UK

T. Lenarz and P. Wriggers (eds.), *Biomedical Technology*, Lecture Notes
in Applied and Computational Mechanics 74, DOI 10.1007/978-3-319-10981-7_7

anatomical structures, thus providing anchoring and mechanical stability to the whole device. TAV stents can be classified in two main categories: balloon-expandable and self-expandable. The first category refers to stents made of elasto-plastic alloys (e.g. stainless steel): in order to be implanted, balloon-expandable TAVs are crimped onto the delivering catheter which is equipped with an elastic deflated balloon; once in the aortic position, TAV deployment is obtained by inflating the balloon. In this phase, the stent is plastically deformed to an expanded configuration allowing TAV positioning within the aortic root (AR). On the other hand, self-expandable stents are made of super-elastic alloys (e.g. Nitinol) and do not need balloon inflation to be deployed: in this case, TAV is crimped onto the delivering catheter by means of a constraining sheath, that is simply retrieved to allow the expansion of the stent to its undeformed configuration.

Since the first procedure, more than 50,000 TAV have been implanted worldwide and TAV indications enlarged to lower risk patients: the randomized PARTNER I trial revealed the non-inferiority of TAV with respect to standard surgery in high risk patients and the superiority of TAV in inoperable patients if compared to alternative therapies (e.g. valvuloplasty, drug therapies) [2, 3]. Recently, a further randomized trial, PARTNER II, has been designed to compare TAV with SAVR in low risk patients.

Despite the promising results and the continuous improvements on the outcomes, TAV application is still limited [4]: the cost-effectiveness of TAV with respect to standard approaches is under an open debate and the evidence that TAV should be considered a preferable procedure in moderate and high risk patients has not been demonstrated. For this reason, many cardiovascular centers tend to prefer SAVR when not contraindicated.

One reason of this trend resides on the fact that up to now clinical trials and registries have been focused on elderly patients with comorbidities, while it is envisioned that the potential benefits of TAV could stand out in less critical patients [5]; a second reason could be related to the current limited knowledge on TAV biomechanics: a deeper understanding of phenomenological aspects related to TAV implantation could support a better definition of clinical indications and help to optimize the procedure, as well as the design of TAV devices [6]. Indeed, many complications affect TAV performances: among them, vascular injuries, stroke events, complete heart blocks requiring pacemaker implantation and prosthesis insufficiency are the most frequent [7]. It should be noted that TAV implantation is performed in a complex anatomical site and without excising the pathological native valve, which is the biological structure that mainly interacts with the prosthesis; it has been demonstrated that the presence of the calcifications on the native AV leaflets affects TAV performances, mostly in terms of TAV insufficiency and heart block events [8–10]: these clinical observations suggest the importance to accounting for the degenerative condition of the native AV in clinical investigations.

Recently, finite element (FE) models have been successfully employed as a tool to investigate device implantation behaviour, giving complementary information to clinicians and contributing to address device design and the planning of therapies [11, 12]. As regards TAV, FE analyses have been performed, mostly with the aim of

assessing TAV outcomes in patient-specific models. Capelli and collaborators [13] simulated the implantation of a balloon-expandable TAV stent in five patient-specific models and investigated the effects related to different positions of the device. They also simulated a valve-in-valve procedure in a model where a previous bioprosthetic valve was present in the implantation site. The authors were able to study the stent configuration following implantation, its interaction with the surrounding structures and the mechanical loadings in the struts of the TAV stent.
A similar study was carried out by Wang and collaborators [14], in which the authors also modeled focal calcifications on the AV.
More recently, the work published by Auricchio and co-authors [15] introduced TAV leaflets in the FE analyses of TAV: after simulating the stent deployment in a patient-specific model of the AR, they mapped the TAV leaflet model on the implanted stent configuration and was able to simulate the valve closure, thus investigating also its diastolic behaviour.
While the two former studies have the limitation to focus only on TAV stent implantation and do not investigate the behaviour of TAV after implantation and during the cardiac cycle, the latter has the advantage to partially study this aspect by introducing the TAV leaflet model. Conversely, the work made by Auricchio and colleagues neglects the native AV that, as stated before, is the main structure interacting with the TAV device.

In this paper, we present a FE study in which both the implantation and function of a balloon-expandable TAV device has been investigated; moreover, in the analyses, the presence of the AV leaflet calcifications has been modeled and their influence on TAV outcomes has been evaluated by modeling different stenotic AVs.

2 Methods

In this work, FE analyses of the implantation and function of a balloon-expandable TAV in AR models were carried out using the explicit solver of LS-DYNA© v. 971 (LSTC, Livermore, CA, USA).
The methods will be divided in two main sections:

- the first section will concern the realization of the AR models and the analysis of their dynamic behaviour;
- the second section will concern the implantation simulation and the dynamic function of the TAV within the AR models.

2.1 Aortic Root Models Definition and Dynamics

In order to investigate the influence of AV calcifications, three different AR models were considered:

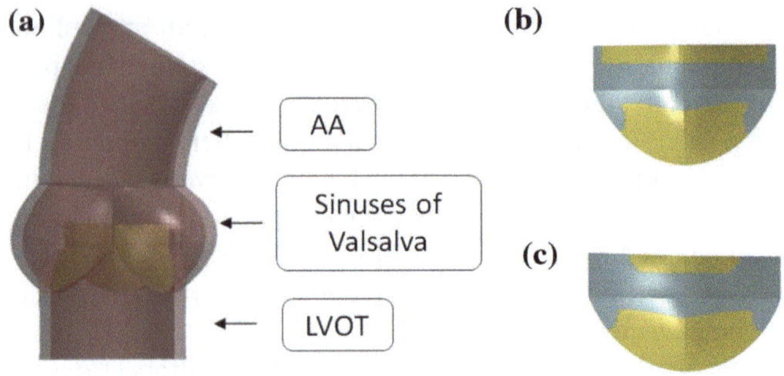

Fig. 1 **a** Representation of the healthy AR model (model H) [16], **b** distribution of the calcific elements on model P1 and **c** model P2 (calcifications are shown in *cyan*)

- a healthy model (model H);
- a first pathological model (model P1) obtained by introducing a typically observed distribution of calcifications on model H;
- a second pathological model (model P2) obtained by extending the calcifications of model P1 on the commissural zones of the AV leaflets.

Model H was taken from a previously published work by Sturla and colleagues [16]: it consists of a paradigmatic physiological AR model based on MRI in vivo data of 8 healthy subjects. The model is shown in Fig. 1a and includes: the AR wall, characterized by the left ventricle outflow tract (LVOT), the sinuses of Valsalva and the first tract of the ascending aorta (AA), and the trileaflet AV. The latter was discretized with 4-node Belytschko–Tsay shell elements [17] with 3 integration points along the out-of-shell direction in order to account for shell bending, while the AR wall was discretized with reduced-integration hexahedral elements in order to limit computational costs.

As concerns material properties, a linear elastic model was assigned to the AR wall with a young modulus of 2 MPa, a Poisson coefficient of 0.45 and a density of 1,100 kg/m^3 [18], while the AV leaflets were modeled with a nearly incompressible transversely isotropic hyperelastic material. Such a model is based on the strain energy function proposed by Weiss and colleagues [17] and was chosen in order to account for the anisotropic mechanical response of the AV leaflets which is stiffer in the direction of the collagen fibers (i.e. the circumferential direction) as shown in the work done by Billiar and Sacks [19].

Model P1 and P2 differ from model H only for the modeling of the AV leaflets.

In order to model AV calcifications, in model P1 a layer of stiffer shell elements (representing the calcifications) has been superimposed onto the healthy AV shell elements: the distribution of the calcific elements was chosen in order to resemble typically observed calcifications on excised stenotic AVs [20]. Model P2 was then obtained by extending the distribution of the calcific elements in the commissural zones of the leaflet. In Fig. 1b, c the distribution of the calcific elements on one of the AV leaflet is shown for model P1 and P2.

As regards the mechanical behaviour of the calcific elements, an isotropic hyperelastic constitutive model based on the Ogden strain energy function [17] has been chosen, by fitting the mechanical response of calcific atherosclerotic tissue published by Loree and colleagues [21].

In order to evaluate the dynamic behaviour of these models and compare the pathological models to model H, structural dynamic simulations of the cardiac cycle were performed, as described by Sturla [16]. Briefly, standard physiological time-dependent aortic and ventricular pressures were applied to the inner surface of the AR wall, after applying a preload of 85 mmHg: the aortic pressure was applied to the structures downstream the AV, while the ventricular pressure was imposed on the LVOT inner surface. On the AV leaflet surface the difference between the ventricular and aortic pressure was applied.

2.2 *TAV Simulations*

In what follows TAV models and simulations will be described in details.

TAV Model. The balloon-expandable Edwards Sapien® size 26 mm (Edwards Lifesciences Corporations, Irvine, CA, USA) was considered. This device is made of a stainless steel (AISI 316 LVM) stent, a bovine pericardium valve and a fabric skirt. The TAV stent model, previously described in the work of Tzamtzis and colleagues [22], was discretized with 169,000 reduced integration tetrahedral elements using the software Gambit® (Ansys Inc., Canonsburg, PA, USA). A finer mesh was realized in the regions of the stent that experience higher level of mechanical stresses during the crimping and deployment phases (e.g. at the joints of the beams). A bilinear elasto-plastic model based on the Von Mises yielding criterion was used to define the stent mechanical properties: the model parameters were obtained from experimental mechanical characterization ($E = 293$ GPa, $\nu = 0.3, \sigma_y = 340$ MPa, $E_t = 715$ MPa) [22].

The leaflet model was discretized with 10,800 linear 3-node shell elements [17]. An isotropic hyperelastic Ogden constitutive model was chosen and the strain energy function parameters were obtained by fitting uniaxial tensile tests performed on bovine pericardium specimens at 37° C. In Fig. 2a the CAD model of the entire TAV device is shown.

TAV Simulations. The implantation and function of the TAV model was simulated through a multi-step procedure, summarized as follows:

1. Crimping simulation of the TAV stent
2. Deployment simulation of the stent within the three AR models
3. Positioning of the TAV leaflet model within the three post-implantation stent configurations
4. TAV function simulation throughout the cardiac cycle

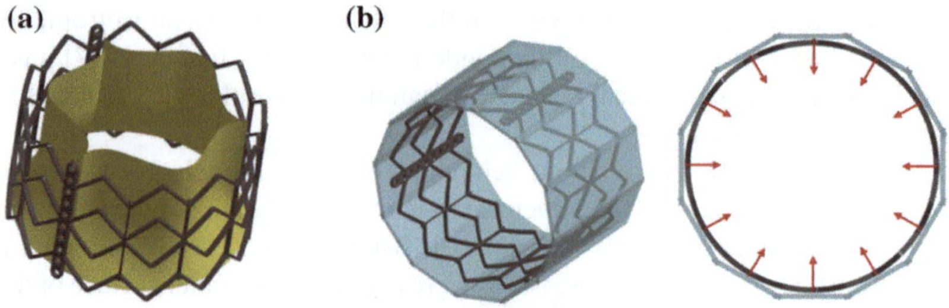

Fig. 2 **a** CAD model of the Edwards Sapien® device **b** Isometric (*left*) and top (*right*) view of the crimping simulation model. In *cyan* the 12 rigid planes are shown and the radial directions of the motion imposed to them is indicated by *red arrows*

Crimping simulation. The crimping simulation was implemented by applying a radial translation to 12 rigid planes placed around the TAV stent (Fig. 2b) and compressing it to a final external radius of 4.7 mm, replicating the real procedure. A frictionless penalty-based contact was used to manage the interaction between the stent and the planes. At the end of the crimping phase a further step was performed, in which the rigid planes were removed and the stent recovered part of its elastic deformation (this phase is called *recoil*).

Deployment simulations. The stent configuration obtained after the crimping and recoil simulation was imported together with its residual stress and strain fields in subsequent analyses, in which the stent was deployed within the three different AR models H, P1and P2 at their systolic peak configurations (Fig. 3). The deployment simulations were performed by applying a uniform pressure (that gradually increased from 0 to 600 kPa) on the inner surface of the stent. As for the crimping simulation, penalty-based contacts were used to manage the interaction between the stent and the native AV, the stent and the AR wall, as well as the AV and the AR wall. Concerning the contacts involving the stent and the biological structures, a formulation that computes the contact forces based on the nodal mass instead of the element stiffness was used, so to cope with the large difference in mechanical properties between soft tissues and steel. In order to limit inertial oscillations due to the choice of a fully-transient formulation, a damping contribution was set: the proper damping coefficient was computed from a non-damped simulation that allowed to obtain the critical frequency of the system.

TAV leaflet positioning simulations. The TAV leaflet model was then positioned in the post-implantation stent configurations through a step in which a non uniform displacement field was applied to the nodes lying on the outer boundaries of the leaflets (except the nodes belonging to the free edges).

TV dynamic simulations. Once the implanted leaflet configurations in model H, P1 and P2 were obtained, dynamic simulations of TAV throughout the cardiac cycle were performed, by applying a physiological trans-valvular pressure on the leaflet surface. In this phase, the nodes belonging to the outer boundaries of the leaflets

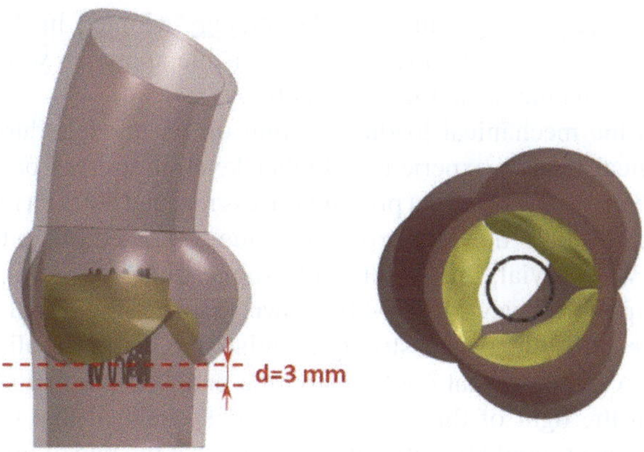

Fig. 3 Side (*left*) and bottom (*right*) view of the initial configuration of the deployment simulation. The distance set between the annulus and the base of the TAV stent is reported

were fully constrained, according to the approximation that the stent does not move during the cardiac cycle.

3 Results

3.1 Dynamics of the Aortic Root Models

The most interesting result obtained from the dynamic simulations of the AR models concerns the Aortic Orifice Area (AOA), whose systolic peak values were 2.3, 1.4 and 1.3 cm^2 in models H, P1 and P2, respectively. In Fig. 4 the AOA at systolic peak are depicted.

Models P1 and P2 showed a significant lowering in AOA values if compared to model H (−39 % and −43 % at systolic peak, respectively). Interestingly, the obtained values are comparable to those of stenotic AV measured by Doppler

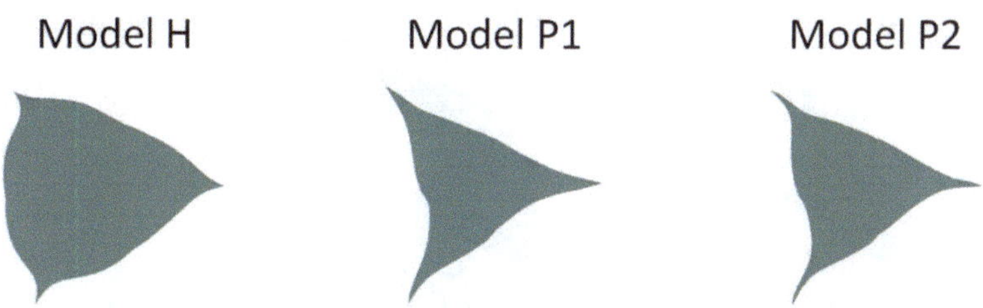

Fig. 4 Systolic peak orifice area of models H, P1 and P2

echocardiography [23], suggesting that the strategy adopted in the present work to model AV calcifications could represent a feasible approach when investigating the macroscopic dynamic behaviour of stenotic AVs.

As concerns the mechanical loadings acting on AV models during the cardiac cycle, pathological models experienced higher level of stresses particularly in the belly region, where the average first principal stress reached 500 kPa in models P1 and P2 and exceeded by 11 % the corresponding value computed in model H (450 kPa). Although somewhat trivial, this result could suggest interesting implications from a pathogenetic point of view: it is well known that the origin and progression of the degenerative aortic stenosis is strictly correlated to higher localized mechanical stresses on the AV leaflets that could lesion the endothelium and start the fibrotic process [24]; in the light of this result it can be speculated that the degenerative process of AV stenosis could be self-enhanced and that the rate of progression of the lesion could increase after the onset of the pathology.

3.2 TAV Simulations

Concerning TAV simulations, in the stent crimping phase mechanical stresses were higher at the joints of the beams where plastic hinges formed (Fig. 5a). In those regions, average Von Mises stress and plastic deformation reached 425 MPa and 13.8 %, respectively. After the recoil phase, the stent recovered part of the elastic deformation, but because of the non-uniform plasticization, a residual auto-equilibrated stress field remained (Fig. 5b). As a result of the elastic recoil, the external radius of the stent increased from 4.7 to 4.9 mm when the rigid planes were removed, which corresponds to a percentage recoil of 4.3 %.

In Fig. 6 the post-implantation configuration of model P2 is reported as an example of the results obtained from the deployment simulations.

The stent configuration in the three models was characterized in terms of average external radius and distal orifice area (Table 1).

In the two pathological models a lower degree of stent expansion was achieved. Results were in agreement with in vivo CT data [25] that reported an average external

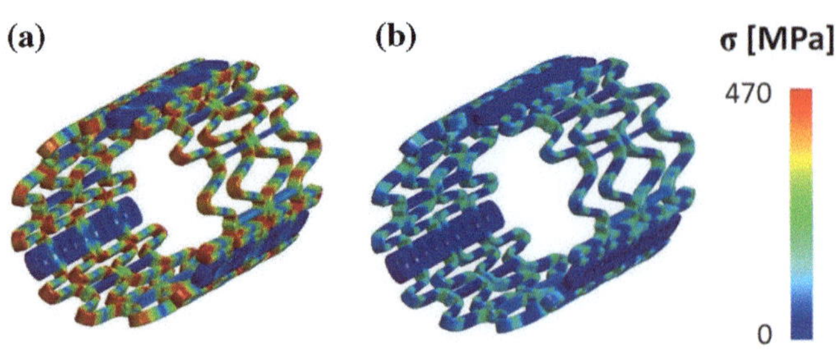

Fig. 5 **a** Von Mises stress in the stent in the end of the crimping phase and **b** of the recoil phase

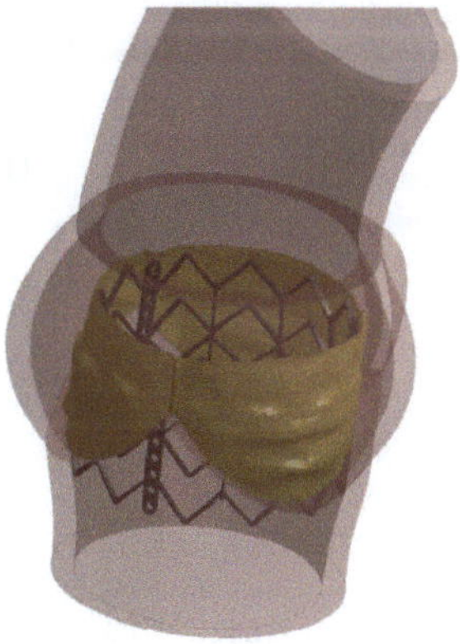

Fig. 6 Post-implantation configuration of model P2

Table 1 Average external radius and distal orifice area of the post-implanted stent configuration in the three models (H, P1, P2)

	Model H	Model P1	Model P2
Average external radius (mm)	13.1	12.6	12.6
Distal orifice area (cm^2)	5.1	4.3	4.0

Table 2 Contact forces computed on the LVOT and on the AV in the three models (H, P1, P2)

	Contact forces (N)	
	LVOT	AV
Model H	12.6	9.7
Model P1	6.2	45.3
Model P2	6.6	50.8

radius of 12.9 ± 0.5 mm for the implanted Edwards Sapien® device and also in accordance to in vivo CT measured distal orifice area [26] that reported average values of $4.0 \pm 0.5\,cm^2$. The distal orifice area of models P1 and P2 is also similar to the values reported by Capelli and colleagues in their computational study [13].

The distortion of the post implanted stent configurations was investigated, by computing the normalized radius along the axial direction of the stent. The results obtained in the three models are shown in Fig. 7a. Model P2 showed the higher degree of distortion (Fig. 7b): in this model the stent configuration was distally less expanded, where the stent interacts with the native AV at the commissural level. In the light of these results it can be claimed that the degree and location of calcifications on the stenotic AV could affect the outcomes of TAV implantation, potentially interfering with the expansion of the TAV stent.

The interaction between the stent and the anatomical structures was also analyzed: the interaction forces were computed separately on the LVOT and on the AV in the three models (Table 2).

A different trend was observed between the healthy and the pathological models: in the former, the interaction mainly affected the LVOT, while the forces acting

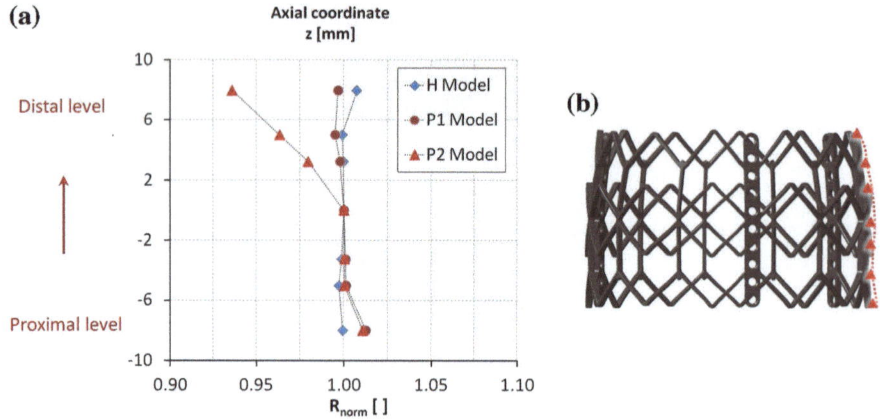

Fig. 7 **a** Normalized stent external radius along the axial coordinate of the stent for the three models (H, P1, P2). **b** Post-implantation configuration of the stent in model P2 where the distorted longitudinal profile is shown

on the AV were lower; conversely, in both the pathological models (P1 and P2) the interaction forces were lower on the LVOT and increased on the AV, reaching higher values in model P2 were wider calcifications were present. The contact forces computed on the LVOT were lower than those reported by Wang and collaborators [14], who found values of 25 N: this difference can be justified by the more proximal position of the stent simulated by Wang. The biomechanical interaction between the stent and the anatomical structures is an important aspect with respect to the critical issue of the impairment of the cardiac tissue, thus explaining the relevance of these results.

In Fig. 8a the TAV leaflet model after the positioning simulation in model P2 is depicted and TAV dynamics at the early systole and during diastole are shown (Fig. 8b, c).

All the three models reported physiological AOA values with systolic peak values of 3.1, 3.1 and 2.7 cm^2 in model H, P1 and P2 respectively.

Moreover, those values were in agreement with in vivo measurements obtained with Doppler echocardiography [26]. It is interesting to note that in model P2 AOA was 13 % lower than in model P1, suggesting another potential implication of the distribution of AV calcifications on TAV performances. Negligible differences were observed in terms of mechanical stresses and strains on the TAV leaflets of the three models, where the average value of the first principal stress was 148, 145, 150 kPa in model H, P1 and P2, respectively. These values are lower than those found by Auricchio and colleagues [15], but this difference can be explained by the different constitutive models applied to the TAV leaflets: while in the present study a non linear hyperelastic mechanical behaviour was used, in their work Auricchio and co-authors approximated bovine pericardium as a linear elastic material.

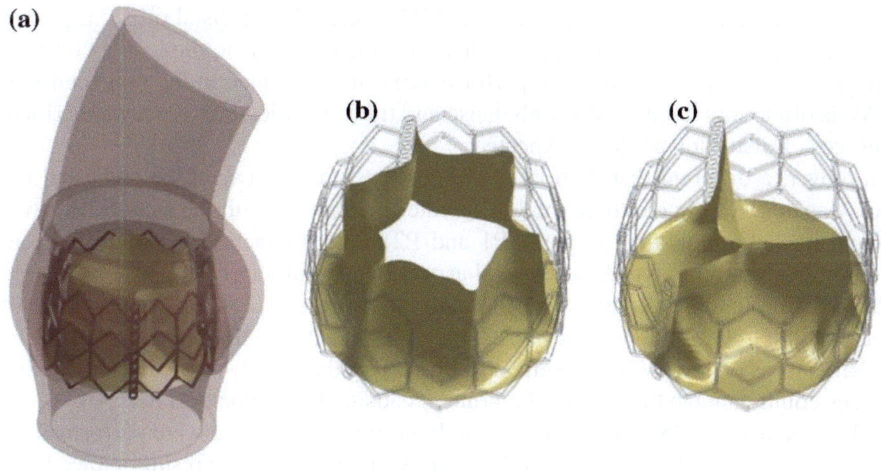

Fig. 8 **a** TAV leaflet in model P2 after positioning simulation, **b** TAV model at early systole and at **c** late diastole

4 Discussion

TAV implantation are currently considered a feasible procedure in high risk and inoperable patients affected by severe aortic stenosis [27]. The current indications still limit TAV to a minority of clinical cases, while it can be claimed that moderate and low risk patients could also benefit from TAV advantages (i.e. avoid general anesthesia and the trauma deriving from SAVR, possibility to perform valve-in-valve implantation) [5]. The assessment of TAV in lower risk patients is still far to come, but the potentials of this endovascular approach are promising. In this context computational methods and tools could play an important role in supporting clinical investigations, as well as the process of device design, with the aim of improving TAV performances and overcome its current limitations and critical aspects.

Having this in mind, in this paper a FE approach for the simulation of both the implantation and function of a TAV model was described: the approach was applied to three MRI-based AR models in which the degree of AV stenosis differed, ranging from absence of calcific stenosis (model H), typically observed early calcifications (model P1) and extended calcifications on the commissural zones of the AV (model P2).

To the best of our knowledge, the proposed modeling approach is the first one that accounts for both native AV leaflets and TAV leaflets and that analyses the post-implantation scenario throughout the cardiac cycle. Moreover, in the developed FEMs the anisotropic and non linear behaviour of the native AV, as well as the non linear mechanical response of the TAV pericardium leaflets was included.

The analyses allowed to investigate how the degree and location of AV calcifications could affect TAV performances: indeed, a distortion in the post-implanted

stent configuration was observed in model P2 where a lower distal orifice area was reached. This result suggests that calcifications located close to the leaflet commissures could potentially affect the performances of TAV, since in the real scenarios TAV is often implanted in AVs with dense and thick calcifications [28], were critical degrees of distortion could be expected.
Moreover, a different distribution of the contact forces between the stent and the AR was observed in the three models: greater interaction with the stenotic AV was found in the pathological models (P1 and P2). The interaction between the stent and the surrounding tissues is a crucial aspect to take into account when evaluating the possible damages induced on the cardiac structures [29]. From this point of view, in the light of the present results, it can be speculated that the presence of the calcifications on the AV has the effect of protecting the LVOT, where the stent radial forces could damage the left bundle brunch, causing heart block events [30].

In conclusion, TAV simulations throughout the cardiac cycle were performed allowing to evaluate the dynamics of TAV following implantation: results showed that in all the three models a physiological behaviour was recovered. Nevertheless, some differences were observed in terms of AOA between the pathological models, suggesting a possible correlation between lower AOA and commissural calcifications.

5 Limitations and Future Perspectives

The present work demonstrated the potential of computational models to investigate TAV implantation outcomes: the results showed how the influence of the calcifications of the native AV may affect the configuration of the implanted device and consequently the global outcomes of the procedure. The findings give insights into the interaction between TAV device and the stenotic AV and the AR, but further improvements and studies are still needed to overcome some limitations.

First of all, the balloon for stent expansion was not modeled in order to reduce the complexity of the analysis and its computational cost: although previous computational studies on percutaneous approaches simulated the deployment of the stent by directly applying a pressure load on its inner surface [11], some authors revealed that this strategy leads to different results when simulating coronary stent deployment [31]. In the future, further efforts will be made in order to introduce the expanding balloon model and to assess how it can affect the results of the performed analyses.

A second limitation is related to the AR models: despite the paradigmatic model proposed by Sturla [16] has been obtained from in vivo data, it is based on the hypothesis that the shape of the aortic annulus is circular. Many clinical studies revealed that the eccentricity of the aortic annulus (that is frequent both in healthy and pathological AR anatomies) has a great influence on the outcomes of TAV, mainly on TAV insufficiency [32]. The application of the developed computational models to different and more realistic AR models could significantly improve the trustworthiness of the results and will be pursued in the next future.

To this aim a more realistic modeling of the AV calcifications is also needed, since in this study ideal and simplified models of AV stenosis were realized: this limitation can be difficult to tackle due to the lack in the literature of morphological and mechanical characterization of calcific AV leaflets.

Finally, a validation of the described FE models is needed to assess the reliability of the results: at this aim, a comparison with an experimental bench test equipped with a high-frequency acquisition camera in terms of macroscopic quantities should be done.

In conclusion, a complete FE approach has been presented that allow the simulation of the different phases of TAV implantation and its function immediately following the procedure. The possibility to investigate the biomechanics and the dynamic behaviour of TAV has been demonstrated. The developed FEMs could be applied to more realistic scenarios to support and direct clinical decisions.

References

1. Cribier, A., Eltchaninoff, H., Bash, A., Borenstein, N., Tron, C., Bauer, F., Derumeaux, G., Anselme, F., Laborde, F., Leon, M.B.: Percutaneous transcatheter implantation of an aortic valve prosthesis for calcific aortic stenosis. Circulation **106**, 3006–3008 (2002)
2. Rodés-Cabau, J.: Transcatheter aortic valve implantation: current and future approaches. Nat. Rev. (Cardiology) **9**, 15–29 (2011)
3. Svensson, L.G., Tuzcu, M., Kapadia, S., Blackstone, E.H., Roselli, E.E., Gillinov, A.M., Sabik, J.F.I., Lytle, B.W.: A comprehensive review of the PARTNER trial. J. Thorac. Cardiovasc. Surg. **145**(3), S11–S16 (2013)
4. Rosengart, T.K., Feldman, T., Borger, M.A., Vassiliades, T.A., Gillinov, A.M., Hoercher, K.J., Vahanian, A., Bonow, R.O., O'Neill, W.: Percutaneous and minimally invasive valve procedures. A scientific statement from th American Heart Association. Circulation **117**(13), 1750–1767 (2008)
5. Vahanian, A., Himberta, D., Brochet, E., Depoix, J., Iung, B., Nataf, P.: Transcatheter aortic valve implantation. Arch. Cardiovasc. Dis. **105**, 181–186 (2012)
6. Fann, J.I., Chronos, N., Rowe, S.J., Michiels, R., Lyons, B.E., Leon, M.B., Kaplan, A.V.: Evolving strategies for the treatment of valvular heart disease: preclinical and clinical pathways for percutaneous aortic valve replacement. Catheter. Cardiovasc. Interv. **71**, 434–440 (2008)
7. Webb, J.G., Wood, D.A.: Current status of transcatheter aortic valve replacement. J. Am. Coll. Cardiol. **60**(6), 483–492 (2012)
8. Ewe, S.H., Ng, A.C., Schuijf, J.d, van der Kley, F., Colli, A., Palmen, M., de Weger, A., Marsan, N.A., Bax, J.J., Delgado, V.: Location and severity of aortic valve calcium and implications for aortic regurgitation after transcatheter aortic valve implantation. Am. J. Cardiol. **108**, 1470–1477 (2011)
9. Haensig, M., Lehmkuhl, L., Rastan, A.J., Kempfert, J., Mukherjee, C., Gutberlet, M., Holzhey, D.M., Mohr, F.W.: Aortic valve calcium scoring is a predictor of significant paravalvular aortic insufficiency in transapical-aortic valve implantation. Eur. J. Cardio-Thorac. Surg. **41**, 1234–1241 (2012)
10. Bleiziffer, S., Ruge, H., Horer, J., Hutter, A., Geisbüsch, S., Brockmann, G., Mazzitelli, D., Bauernschmitt, R., Lange, R.: Predictors for new-onset heart block after transcatheter aortic valve implantation. J. Am. Coll. Cardiol. Cardiovasc. Interv. **3**(5), 524–530 (2010)
11. Schievano, S., Taylor, A., Capelli, C., Lurz, P., Nordmeyer, J., Migliavacca, F., Bonhoeffer, P.: Patient specific finite element analysis results in more accurate prediction of stent fractures. J. Biomech. **43**, 687–693 (2010)

12. Capelli, C., Taylor, A.M., Migliavacca, F., Bonhoeffer, P., Schievano, S.: Patient-specific reconstructed anatomies and computer simulations are fundamental for selecting medical device treatment: application to a new percutaneous pulmonary valve. Philos. Trans. R. Soc. **368**, 3027–3038 (2010)
13. Capelli, C., Bosi, G.M., Cerri, E., Nordmeyer, J., Odenwald, T., Bonhoeffer, P., Migliavacca, F., Taylor, A.M., Schievano, S.: Patient-specific simulation of transcatheter aortic valve stent implantation. Med. Biol. Eng. Comput. **50**(2), 183–192 (2012)
14. Wang, Q., Sirois, E., Sun, W.: Patient-specific modeling of biomechanical interaction in transcatheter aortic valve deployment. J. Biomech. **45**(11), 1965–1971 (2012)
15. Auricchio, F., Conti, M., Morganti, S., Reali, A.: Simulation of transcatheter aortic valve implantation: a patient-specific finite element approach. Computer Methods in Biomechanics and Biomedicals Engineering. **17**(12), 1347–1357 (2014)
16. Sturla, F., Votta, E., Stevanella, M., Conti, C.A., Redaelli, A.: Impact of modeling fluid-structure interaction in the computational analysis of aortic root biomechanics. Med. Eng. Phys. **35**(12), 1721–1730 (2013)
17. Hallquist, J.O.: LS-DYNA Theoretical Manual. Livermore Software Technology Corporation, California (2000)
18. Conti, C.A., Votta, E., Della Corte, A., Del Viscovo, L., Bancone, C., Cotrufo, M., Redaelli, A.: Dynamic finite element analysis of the aortic root from MRI-derived parameters. Med. Eng. Phys. **32**, 212–221 (2010)
19. Billiar, K.L., Sacks, M.S.: Biaxial and mechanical properties of the natural and glutaraldehyde treated aortic valve cusp—Part I: experimental results. J. Biomech. Eng. **122**, 23–30 (2000)
20. Thubrikar, M.J., Aouad, J., Nolan, S.P.: Patterns of calcific deposits in operatively excised stenotic or purely regurgitant aortic valve and their relation to mechanical stress. Valv. Heart Dis. **58**, 304–308 (1986)
21. Loree, H.M., Grodzinsky, A.J., Park, S.Y., Gibson, L.J., Lee, R.T.: Static circumferential tangential modulus of human atherosclerotic tissue. J. Biomech. **27**(2), 195–204 (1994)
22. Tzamtzis, S., Viquerat, J., Yap, J., Mullen, M.J., Burriesci, G.: Numerical analysis of the radial force produced by the Medtronic-CoreValve and Edwards-SAPIEN after transcatheter aortic valve implantation (TAVI). Med. Eng. Phys. **35**(1), 125–130 (2013)
23. Razzolini, R., Longhi, S., Tarantini, G., Rizzo, S., Napodano, M., Abate, E., Fraccaro, C., Thiene, G., Iliceto, S., Gerosa, G., Basso, C.: Relation of aortic valve weight to severity of aortic stenosis. Am. J. Cardiol. **107**(5), 741–746 (2011)
24. Dweck, M.R., Boon, N.A., Newby, D.E.: Calcific aortic stenosis. A disease of the valve and the myocardium. J. Am. Coll. Cardiol. **60**(19), 1854–1863 (2012)
25. Willson, A.B., Webb, J.G., Gurvitch, R., Wood, D.A., Toggweiler, S., Binder, R., Freeman, M., Madden, M., Hague, C., Leipsic, J.: Structural integrity of balloon-expandable stents after transcatheter aortic valve implantation. JACC: Cardiovasc. Interv. **5**(5), 525–532 (2012)
26. Delgado, V., Ng, A.C., van de Veire, N.R., van der Kley, F., Schuijf, J.D., Tops, L.F., de Weger, A., Tavilla, G., de Rood, A., Kroft, L.J., Schalij, M.J., Bax, J.J.: Transcatheter aortic valve implantation: role of multidetector row computer tomography to evaluate prosthesis positioning and deployment in relation to valve function. Eur. Heart J. **31**, 1114–1123 (2010)
27. Al-Lamée, R., Godino, C., Colombo, A.: Transcatheter aortic valve implantation. Current principles of patient and technique selection and future perspectives. Circ. Cardiovasc. Interv. **4**, 385–387 (2011)
28. Zegdi, R., Ciobotaru, V., Noghin, M., Sleilaty, G., Lafont, A., Latrémouille, C., Deloche, A., Fabiani, J.: Is it reasonable to treat all calcified stenotic aortic valves with a valved stent? J. Am. Coll. Cardiol. **51**(5), 579–584 (2008)
29. Piazza, N., de Jaegere, P., Schultz, C., Becker, A., Serruys, P., Anderson, R.: Anatomhy of the aortic valvar complex and its implications of transcatheter implantation of the aortic valve. Catheter. Cardiovasc. Interv. **1**, 74–81 (2008)
30. Léon, M.B., Piazza, N., Nikolski, E., Blackstone, E.H., Cutlip, D.E., Kappetein, A.P., Krucoff, M.W., Mack, M., Mehran, R., Miller, C., Morel, M., Petersen, J., Popma, J.J., Takkemberg, J.M., Vahanian, A., van Es, G., Vranckx, P., Webb, J.G., Windecker, S., Serruys, P.W.: Standardized

endpoint definitions for transcatheter aortic valve implantation clinical trials. J. Am. Coll. Cardiol. **57**(3), 253–269 (2011)
31. Gervaso, F., Capelli, C., Petrini, L., Lattanzio, S., Di Virgilio, L., Migliavacca, F.: On the effects of different strategies in modelling balloon-expandable stenting by means of finite element method. J. Biomech. **41**, 1206–1212 (2008)
32. Pontone, G., Andreini, D., Bartorelli, A.L., Bertella, E., Cortinovis, S., Mushtaq, S., Annoni, A., Formenti, A., Baggiano, A., Conte, E., Tamborini, G., Muratori, M., Gripari, P., Bovis, F., Veglia, F., Foti, C., Alamanni, F., Ballerini, G., Fiorentini, C., Pepi, M.: Aortic annulus area assessment by multidetector computed tomography for predicting paravalvular regurgitation in patients undergoing balloon-expandable transcatheter aortic valve implantation: a comparison with TTE and TEE. Am. Heart J. **164**(4), 576–584 (2012)

Repair of Mitral Valve Prolapse Through ePTFE Neochordae: A Finite Element Approach From CMR

Francesco Sturla, Francesco Onorati, Emiliano Votta, Marco Stevanella, Aldo D. Milano, Konstantinos Pechlivanidis, Giovanni Puppini, Alberto Redaelli and Giuseppe Faggian

Abstract Patient-specific finite element (FE) modeling is largely used to quantify mitral valve (MV) biomechanics associated to pathological and post-surgical conditions. We used this approach, integrated with non-invasive cardiac magnetic resonance (CMR) imaging data, to numerically perform the repair of the isolated mitral valve leaflet prolapse through expanded-polytetrafluoroethylene (ePTFE) sutures and quantitatively compare the effects of different techniques of neochordal implantation (NCI). CMR-derived FE models well reproduced MVP-related alterations and were able to assess the efficacy of each repairing technique and its biomechanical effects on MV apparatus; the quantification of biomechanical differences between NCI techniques, especially in terms of both chordal tensions and leaflet stresses redistribution, may impact on the short- and long-term the clinical outcome, potentially opening the way to patient-specific optimization of NCIs and, if extensively and successfully tested, improve surgical planning.

Keywords Mitral valve surgery · Neochordae · Cardiac magnetic resonance · Finite element models

1 Introduction

Isolated mitral valve posterior leaflet prolapse (MVP) represents the most common dysfunction connected with degenerative mitral valve disease [1]: it is clinically defined as the displacement of the posterior mitral leaflet in the left atrium, past the annular plane. Chordal rupture or elongation are recognized as the main causes of MVP whereas annular dilation is almost always an associate finding.

F. Sturla (✉) · F. Onorati · A.D. Milano · K. Pechlivanidis · G. Puppini · G. Faggian
Division of Cardiovascular Surgery, Università degli Studi di Verona, Verona, Italy
e-mail: francesco.sturla@univr.it

F. Sturla · E. Votta · M. Stevanella · A. Redaelli
Department of Electronics, Informatics and Bioengineering, Politecnico di Milano, Milan, Italy
e-mail: francesco.sturla@mail.polimi.it

T. Lenarz and P. Wriggers (eds.), *Biomedical Technology*, Lecture Notes in Applied and Computational Mechanics 74, DOI 10.1007/978-3-319-10981-7_8

Nowadays it can be treated through several conservative surgical techniques based on artificial neochordae implantation (NCI) [2] which allow for respecting rather than resecting the diseased portion of the mitral valve [3]. Replacement of the chordae tendinae by ePTFE sutures was introduced into clinical practice more than 20 years ago [4] and various reports have been published in literature documenting good early and long-term results with these techniques [5, 6].

However, MVP repair through chordal replacement is technically demanding and determining the appropriate length of the ePTFE suture is one the major concerns [7]: therefore a long-standing experience is required in order to properly face the problem and perform a successful repair. Several patient-dependent aspects (i.e. lesion severity, prolapse extension etc.) must be taken into account, and the optimal NCI. approach has to be chosen among the several ones available, which differ by number and configuration of neochordae [8]. This choice should be based on sound biomechanical criteria in order to select the appropriate solution, for each clinical scenario, and optimize the surgical repair through a patient-specific approach to the problem.

Although FE modeling is well recognized as a suitable approach to investigate MV biomechanics [9] only a few literature studies have attempted to virtually reproduce MVP repair through NCI. A pioneer study simulated ePTFE suture replacement of ruptured posterior chordae tendinae on a paradigmatic and symmetrical MV geometry, derived from measures on fresh porcine heart [10]. A more recent computational study moved towards a more realistic MV model, derived from three-dimensional echocardiography, and simulated MVP repair with ePTFE sutures [11]. Nonetheless, further investigations are necessary in order to deeper assess and understand the wide spectrum effects of NCIs on MV biomechanics.

At this aim, a computational FE strategy, based on in vivo CMR data, is presented herein to deeper investigate MV prolapse repair through implantation of ePTFE-neochordae: the developed approach focuses on MV leaflets repositioning, chordal tension redistribution and tissue stresses assessment after MV repair.

2 Methods

An integrated framework was developed combining CMR technology, FE methods and the use of dedicated software packages: the developed analysis allows the quantitative assessment of MV apparatus and the simulation of different NCI techniques as detailed in the following paragraphs.

2.1 CMR Acquisition and Image Segmentation

Cardiac Magnetic Resonance (CMR) was performed on a male patient (75 years old, 83 kg, 168 cm) who underwent surgery for isolated prolapse of mitral posterior P2 scallop due to multiple rupture of marginal chordae. A customized sequence of acquisition was carried out on a 3.0T TX Achieva (Philips Medical System, Irvine,

Calif) machine: cine-images were obtained through 18 long-axis planes, evenly rotated along the left ventricle (LV) axis every 10 degrees, with an isotropic *pixel spacing*, stored in each DICOM image file as the distance between adjacent pixels, of 1.25 mm and a slice thickness of 8 mm. Using a breath-holding modality, 30 cardiac phases were recorded on each plane, with different temporal resolution according to the characteristic R-R interval of each patient.

The use of a CMR rotational sequence of acquisition allowed to completely acquire MV apparatus and dynamically assess its function through out an entire cardiac cycle. The axis of rotation coincided with LV axis and it was defined as the segment connecting LV apex and mitral valve centroid (as schematically reported

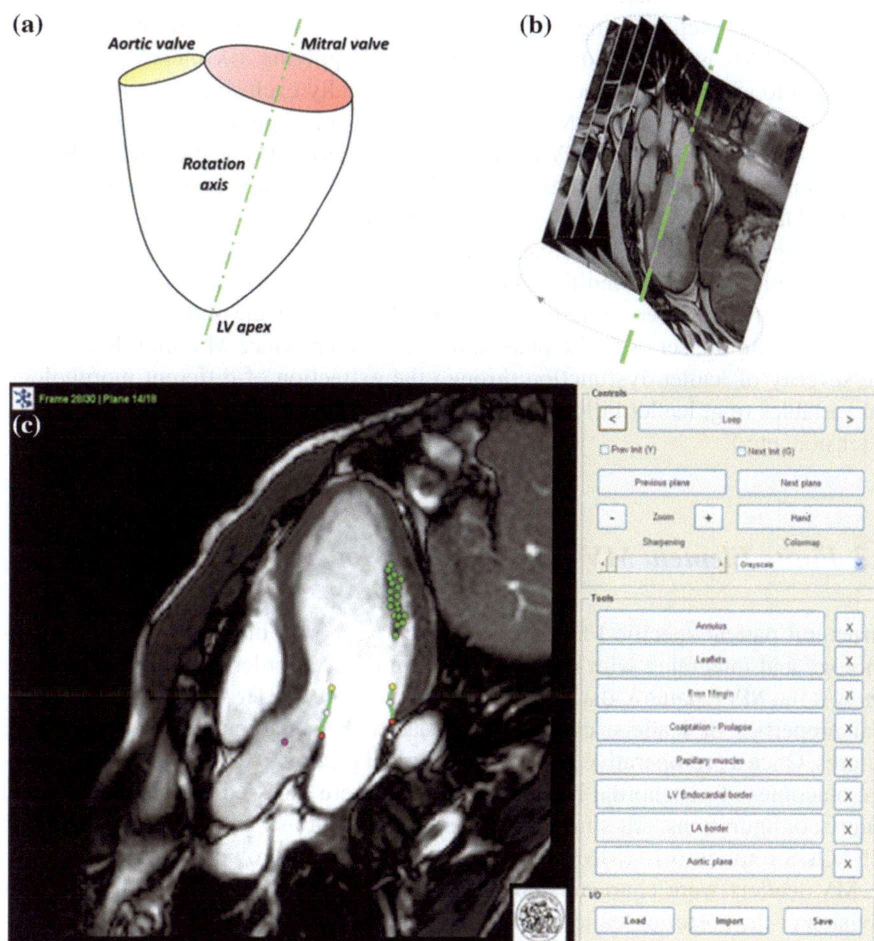

Fig. 1 CMR-acquisition protocol and offline segmentation: **a** positioning of LV long-axis of acquisition; **b** derived rotational sequence of acquisition consisting of 18 long-axis planes; **c** view of the segmentation panel with a tracking example of MV apparatus

in Fig. 1a: both points were identified through auxiliary CMR short-axis planes and LV axis was finally arranged after checking the operation on a supplementary LV long-axis plane. For each acquisition, a set of 540 DICOM images were generated and a total acquisition time of about 10 min was required.

Offline CMR images segmentation was performed using a dedicated software [12] developed in MATLAB (The MathWorks Inc., Natick, Mass). As detailed in Fig. 1c, two points (in red color) were manually selected on MV annulus in each of the acquired plane, two points (in yellow color) were positioned on the free margin of the MV leaflets and each papillary muscle (PM), where visible, was added through a variable number of points (green points). Three-dimensional coordinates were then computed using the information stored in the appropriate DICOM fields.

Mitral annular geometry was automatically generated through a 4th order Fourier approximating function fitting the selected annular points; both anterolateral and posteromedial PMs were positioned in the 3D space applying a *K*-means partitioning algorithm to separate reference points in two mutually exclusive clusters and identifying the centroid of each PM tip. An additional set of points, was positioned on the long-axis profile of each MV leaflet in order to derive MV leaflet profile through a polynomial fitting (green lines), its local inclination (with respect to acquisition axis) and leaflet extension, i.e. the distance between annulus and leaflet free-margin.

The segmentation of CMR images provided MV end-diastolic configuration and the kinematic boundary conditions representing annular and PMs dynamics. In addition, MV leaflet surface was assessable at systolic peak, i.e. the mid-frame of systolic interval within the set of CMR-phases, in order to reproduce MV morphology and the severity of leaflet dysfunction through the extraction of different morphological variables (e.g. leaflet billowing height and volume, regurgitation area, prolapse extension, etc.).

2.2 Finite Element MV Pre-operative Model

Structural patient-specific FE models were derived from the CMR-extracted morphology and integrated with intra-operative findings, in order to reliably reproduce the specific MVP pattern; data from the literature were used to describe the mechanical properties of leaflets, chordae tendineae and, in particular, artificial ePTFE sutures. Once a pre-operative model (with chordal rupture) was simulated, the potential outcomes of neochordal implantation (NCI) were assessed, evaluating different suture configurations. All simulations were carried out using the commercial solver ABAQUS Explicit 6.10 (SIMULIA, Dassault Systèmes, Providence, RI).

MV leaflets were characterized by a single anterior segment and the peculiar division of the posterior leaflet in three different scallops (i.e. P1, P2 and P3, respectively): the end-diastolic geometry was discretized with a homemade software written in Python (Python Software Foundation, Beaverton, OR): the mesh consisted of 3-nodes shell elements (ABAQUS S3R element type) and a mapped scheme of meshing was adopted. The number and the distribution of chordae tendineae were defined according to a paradigmatic scheme [12] derived from ex vivo findings [13, 14] and it included all MV characteristic types of chordae tendinae: marginal (first order), basal (second order) and structural anterior chordae.

As above mentioned, in order to reproduce a reliable pre-operative model, structural FE models were integrated with intra-operative measurements and surgical findings, thus confirming location and extent of the prolapsing region and the exact number and type of ruptured chordae.

All tissues were assumed as homogeneous ($1.1\,g/cm^3$), non-linear and elastic: their behaviour was modeled on the basis of the hyperelasticity theoretical framework in order to account for their large deformations.

MV leaflet response was described through the hyperelastic constitutive model originally proposed by May-Newman et al. [15], which is based on the strain energy function Ψ that depends on two invariant measures of finite strain I_1 and I_4:

$$\Psi(I_1, I_4) = c_0 \left[e^{c_1(I_1-3)^2 + c_2(\sqrt{I_4}-1)^4} - 1 \right] \tag{1}$$

where $I_1 = tr(\boldsymbol{C})$ and $I_4 = \mathbf{a}_0 \bullet \boldsymbol{C} \bullet \mathbf{a}_0 = \lambda^2$, in which $\boldsymbol{C} = \boldsymbol{F} \bullet \boldsymbol{F}^T$ is the right Cauchy-Green tensor, $\mathbf{a}_0$ is the unit vector which defines the preferential direction of the fibers in the material in the undeformed shape and λ is the stretch of the fibers in the $\mathbf{a}_0$ direction. $\boldsymbol{F}$ is the deformation gradient tensor, defined as $\boldsymbol{F} = \boldsymbol{\partial}\mathbf{x}/\boldsymbol{\partial}\mathbf{X}$, i.e. the derivative of the current position with respect to the undeformed position. The constitutive relation was implemented in a user-defined subroutine (VUMAT), as detailed in [12]: the constitutive parameters identified by May-Newman were used [15] and, accordingly, the direction parallel to the annular profile was defined as the preferential direction of the collagen fibers. Moreover, regionally varying thicknesses were assigned to the anterior and posterior MV leaflets as reported in [16].

Native chordae tendineae response was assumed isotropic and described through hyperelastic models available in ABAQUS/Explicit and fitting data from literature [17] through a 2nd order polynomial function for marginal and structural chordae and a 5th order Ogden function for basal chordae.

For each patient-specific model, two simulations were preliminary implemented: the former represented the pre-operative model (*Pre-Op* model) reproducing the clinical scenario of MV lesions and prolapse dysfunction as derived from CMR images; the latter, replicating the same geometry of *Pre-Op* model, represented the physiological MV model (*Phy* model) without prolapse, i.e. with a complete and intact chordal apparatus with respect to pre-operative conditions. For both models, the systolic MV biomechanics was simulated, applying a physiological pressure load to the leaflets, for a total simulation time of one second. As above mentioned, boundary conditions were extracted from CMR-segmentation and reproduced annular kinematics and PMs displacement. A general contact algorithm, available in ABAQUS with a scale-penalty method and a friction coefficient of 0.05 was used to model MV leaflets coaptation [18].

2.3 Neochordal Implantation (NCI)

Expanded polytetrafluoroethylene (ePTFE) CV-5 neochordae were modeled as non-linear springs (ABAQUS SPRINGA elements) with no resistance to compression and a diameter of 350 μm as reported by the manufacturer: force-elongation behavior was derived from literature data [19] consistently with neochordae undeformed length and cross-section, and the characteristic stress-strain curve of the material (Fig. 2a).

Subsequently, different clinical NCI procedures were simulated (as schematically reported in Fig. 2e) and the proper suture length was determined, for each NC configuration, as follows:

1. **single neochorda** (SN) with suture length L_{NC} approximated in millimeters to the distance, measured in *Phy* model at peak systole, between PM tip and the selected point of neochordal insertion along the free margin of the MV scallop;
2. **double neochorda** (DN) made of 2 different sutures arising from the same PM tip and with different lengths L^1_{NC} and L^2_{NC}, separately measured as in SN configuration;

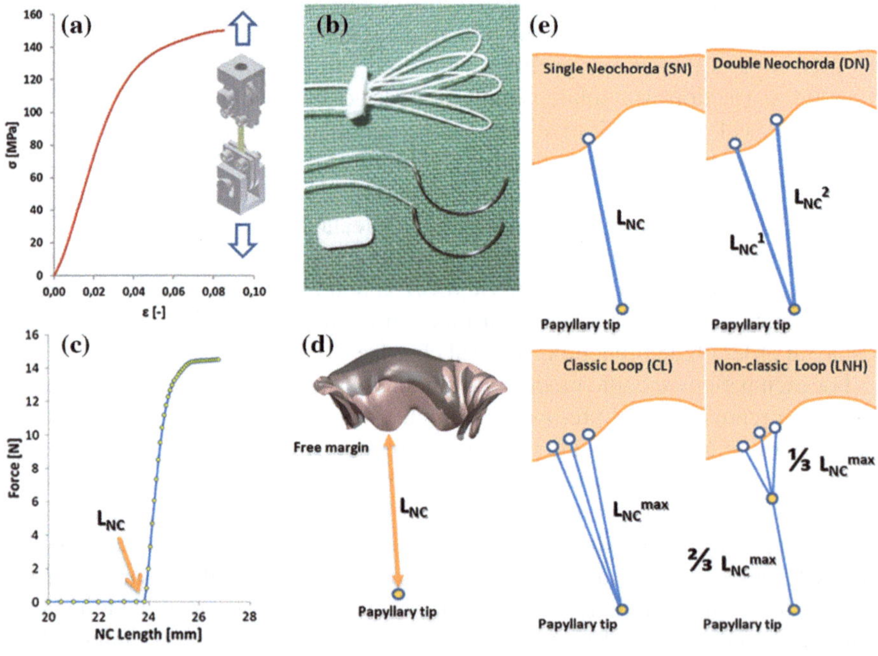

Fig. 2 **a** Stress-strain curve of ePTFE CV-5 sutures as derived from monoaxial traction tests available in literature [19]; **b** typical single and loop configurations of artificial neochordae for MVP repair; **c** computational force-elongation behavior defined for a single ePTFE artificial neochorda with an undeformed length of 24 mm; **d** simulation of the measurement of neochordal length, in the *Phy* model: the distance between the leaflet free margin and the corresponding PM tip was used; **e** schematic representation of the performed NCIs

3. **"classic" loop** (CL) with 3 different neochordae arising from the same PM tip, in which all the 3 neochardae share the same length L_{NC}^{max}, corresponding to the maximal distance between PM tip and leaflet free margin, as measured in *Phy* model;
4. **"non-classic" loop** (LNH) with a common neochordal loop of 2/3 and 3 different neochordae of 1/3 of the entire (PM tip-to-leaflet free margin) length L_{NC}^{max}, determined as in the CL configuration.

3 MV Biomechanics After NCI

The modeling strategy well reproduced MVP-related anatomical alterations as visible from CMR data (Fig. 3a). The simulated *Pre-Op* model is reported at peak systole in Fig. 3. Both the colorbars reported the locally z-displacement of each MV leaflet region with respect to the annular profile: a billowing height higher than 6 mm was reported on the posterior leaflet whereas no prolapse was pre-operatively noticed on the anterior mitral leaflet.

Simulated post-operative systolic function was compared with the corresponding pre-operative simulation (Fig. 2) and assessed in terms of several variables: (i) leaflets coaptation area (CoA), (ii) coaptation length (CoL), (iii) leaflet mechanical stresses, (iv) PMs reaction forces and (v) chordal tensions.

All NCI procedures repositioned the free margin of the prolapsing region of the leaflet below the annular plane and a good leaflet coaptation was obtained between anterior and posterior leaflet restoring a larger triangle of coaptation [20] if compared

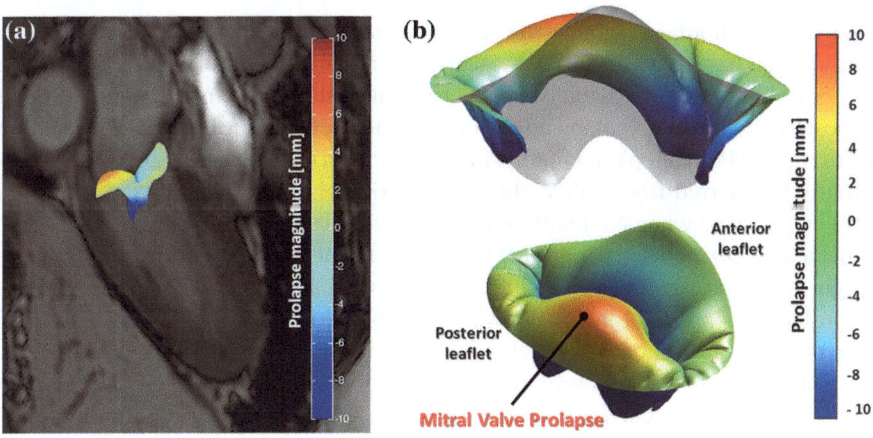

Fig. 3 **a** Simulated *Pre-Op* model reported at peak systole on a long-axis CMR-image; **b** MV pre-operative model with prolapse location on the posterior leaflet. The colorbars report the prolapse magnitude, defined as the distance between leaflet. tissue and the annular profile

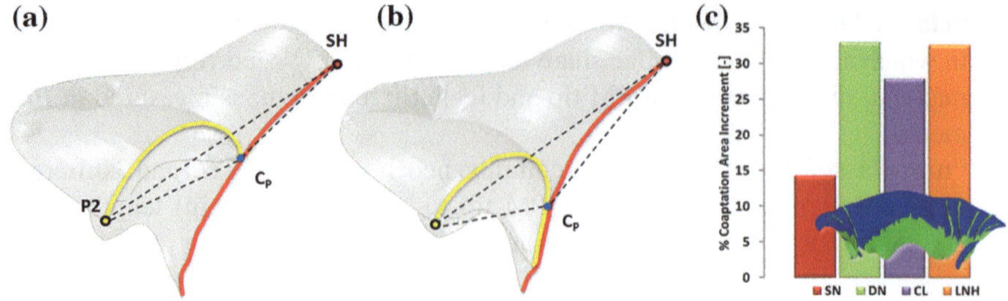

Fig. 4 Representation at peak-systole of the triangle of coaptation in pre-operative conditions (**a**) and after neochordal implantation of a classic loop (**b**); **c** relative percentage CoA increment after different NCIs

with the pre-operative model. The triangle of coaptation (as reported through dotted lines in Fig. 4a, b) is defined connecting the two points sited on the mitral annulus (SH on the septal and P2 on the lateral mitral annulus) with the coaptation point C_P. In addition, all NCI simulations restored a comparable level of CoL (range: 6.2–7.2 mm) with the largest values of CoL obtained when adopting multiple NCI.

As underlined in the histogram of relative CoA increment (Fig. 4c), the use of multiple neochordae, with respect to the *Pre-Op* model, progressively improved the repair in terms of prolapse reduction and coaptation area recovery, passing from SN to multiple NCIs, and determined a wider realignment of the free margin along the P2 prolapsing region.

In all of the NCI models, a slight decrease in PMs reaction forces was noticed: in Fig. 5a the temporal curve of PMs reaction force is reported throughout the entire duration of MV closure simulation, i.e. from initial valve closure to the final systolic peak.

As regards chordal tension, in the prolapsing region, the tension decreased on both marginal and basal native chordae, as starred in red and black color in the histogram (Fig. 5c) where chordal tension is reported at peak systole for all native chordae along the posterior leaflet (the prolapsing region is located between the midline of P2 scallop and the lateral P3 scallop) and for ePTFE sutures. As a consequence, tension load was partially transferred from native chordae to artificial neochordae, this effect being more emphasized when NCI with multiple neochordae was simulated.

In the *Pre-Op* model, a peak of Maximum Principal Stress was noticed on the free-margin of the prolapsing scallop in proximity of a native chorda (Fig. 6a) whereas the prolapsing portion of the leaflet was partially unloaded.

In post-operative models, leaflet stresses along the non-prolapsing scallops were substantially unchanged between pre- and post-operative analyses, regardless of the employed technique. On the contrary, stresses on the prolapsing scallop markedly changed: regardless of the simulated NCI technique, leaflet stresses were reduced in the areas subtended by native chordae located in proximity of the prolapse and increased in the repaired portion of the posterior leaflet since ePTFE sutures restored mechanical tension along the prolapsing region of the leaflet (Fig. 3). Moreover, as

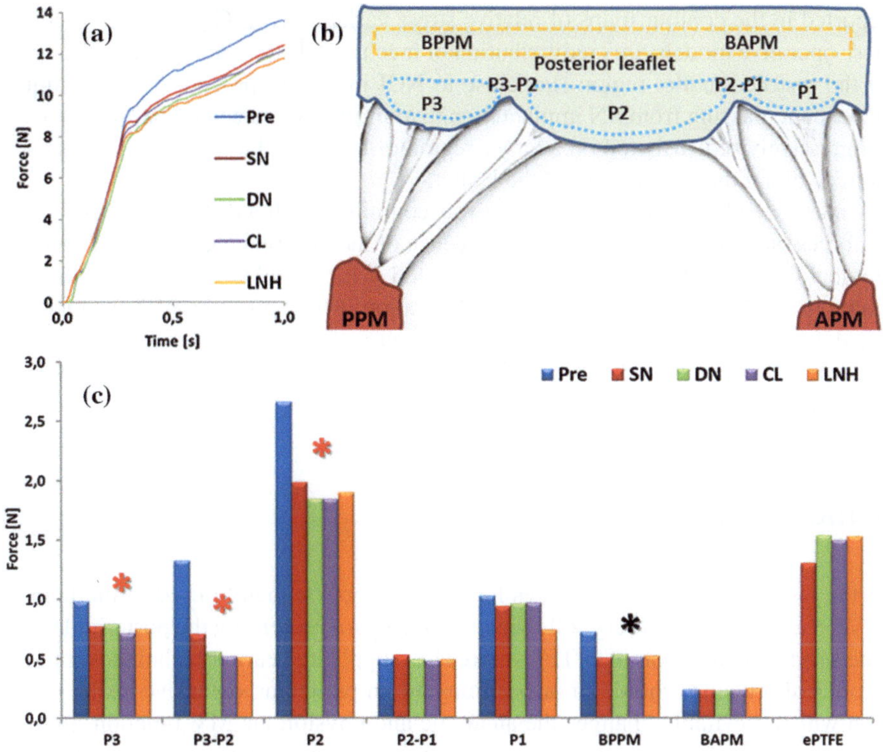

Fig. 5 **a** PMs reaction force for each performed simulation (including *Pre-Op* model); **b** schematic view of the mitral posterior leaflet with the inserting regions of basal chordae arising from the posterior PM (BPPM) and anterior PM (BAPM), and the inserting regions of marginal chordae along the leaflet free-margin of the posterior scallops; **c** native chordae and ePTFE suture forces at peak systole (asterisks highlight a marked reduction in chordal tension, between pre- and post-operative conditions)

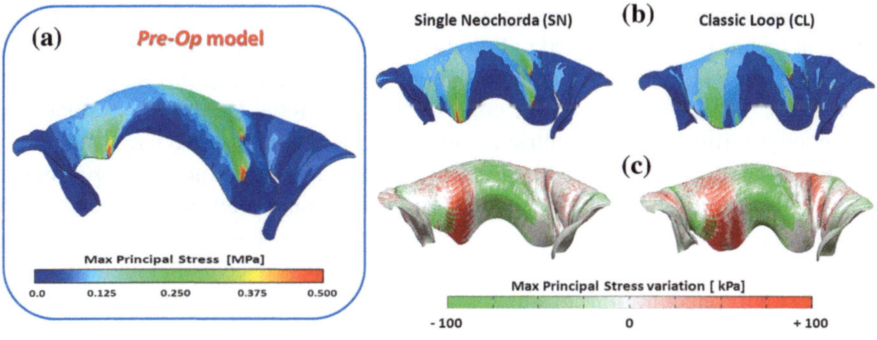

Fig. 6 Mechanical stress on posterior mitral leaflet: **a** Maximum Principal Stress in the *Pre-Op* model at peak systole and, **b**, after different NCIs; **c** stress relative redistribution on posterior leaflet tissue after NCIs

reported in the contour maps of relative stress variation (Fig. 6b), between pre- and post-operative models, this redistribution of mechanical stress along the free-margin of the restored leaflet region (i.e. close to NCI insertion) proved to be technique-dependent: passing from SN implantation to multiple NCI, mechanical stresses were progressively redistributed on a larger region of leaflet and a lower peak stress was noticed along the free margin of the leaflet.

Indeed, leaflet stress redistribution was combined with a different redistribution of forces on ePTFE sutures, as a consequence of the different NCI set-up in terms of number of NCIs. When considering multiple NCI, chordal tension was subdivided on more ePTFE sutures than in SN configuration, where a single ePTFE supported the entire tension. A different load subdivision was noticed for each NCI: 59 and 41 % in DN simulation; 30, 30 and 40 % in CL configuration and 32, 33 and 36 % in LNH simulation.

4 Discussion

In the last decades, ePTFE sutures have been more frequently used in the field of MVP reconstructive surgery following the concept of preserving the posterior leaflet. The surgical innovation of ePTFE sutures has supplanted earlier techniques, such as chordal shortening, chordal transposition and, in particular, extensive leaflet resection which inevitably changes the anatomic and physiologic MV function. Indeed, implantation of ePTFE sutures can restore the proper coaptation area, preserve ventriculo-annular continuity and, minimize leaflet tension [21]. Moreover, a limited tissue reactivity has been noticed: experimental tests have demonstrated that potential infections, due to ePTFE sutures, are comparable with that of other synthetic sutures [22].

Hovewer, from the clinical point of view, MVP repair through ePTFE sutures is still a challenging technique deserving continuous attention over time: the determination of suture length represents a key aspect and it is usually accomplished through different techniques, as reported in literature [23]: the anatomic techniques evaluate the length of ePTFE sutures having as a reference point a non-prolapsing scallop in front of the prolapsing one; on the contrary, the functional techniques evaluate the suture length filling the left ventricle with saline. In this study, the use of an auxiliary *Phy* model, was performed in order to numerically determine the proper suture length reproducing, through the application of a pressure load on MV leaflets, the same operative conditions of saline injection in functional techniques. As a matter of facts, a neochordae that is too short will result in a restricted leaflet mobility (thus inducing high concentrations of stresses on the leaflet); on the contrary, a neo-chordae that is too long will be ineffective in controlling leaflet prolapse [24].

According to the aims of the study, different NCIs were simulated in order to compare the effects of each repair technique on MV apparatus and infer systematic differences between NCIs. Different groups have recently proposed non-standard and innovative NCIs [25–27] in order to improve MV repair and increase the possibility to perform complex MV repair; however, a deeper investigation of the biomechanical effects associated with different NCIs is necessary since, as highlighted, the magnitude and distribution of mechanical stress on MV leaflets proved to be technique-dependent, potentially impacting on the mid- and long- term the surgical outcome of the procedure.

The proposed FEM approach may represent a useful tool of investigation in order to elucidate the biomechanical effects of different NCIs: biomechanical information about MV functional improvements after NCI may help surgeons to determine the proper surgical approach to MVP repair according to each clinical scenario, thus improving and optimizing the clinical application of NCI techniques.

5 Conclusions

The presented approach offers a deeper insight into the biomechanical aspects involved in MVP repair: apart from the "macroscopic" assessment of the coaptation area and length, FE models may provide physicians with a detailed quantification of MV biomechanics after NCIs; highlighted biomechanical differences in post-operative models may potentially impact on the clinical outcome of the procedure and promoting, if extensively and successfully tested, a patient-specific optimization of NCI techniques for MVP repair.

References

1. Iung, B., Baron, G., Butchart, E.G., Delahaye, F., Gohlke-Barwolf, C., Levang, O.W., et al.: A prospective survey of patients with valvular heart disease in Europe: The Euro heart Survey on valvular heart disease. Eur. Heart. J **24**, 1231–1243 (2003)
2. Bortolotti, U., Milano, A.D., Frater, R.W.: Mitral valve repair with artificial chordae: a review of its history, technical details, long-term results, and pathology. Ann. Thorac. Surg **93**, 684–691 (2012)
3. Perier, P., Hohenberger, W., Lakew, F., Batz, G., Urbanski, P., Zacher, M., et al.: Toward a new paradigm for the reconstruction of posterior leaflet prolapse: midterm results of the respect rather than resect approach. Ann. Thorac. Surg. **86**, 718–725 (2008)
4. David, T.E.: Replacement of chordae tendineae with expanded polytetrafluoroethylene sutures. J. Card. Surg. **4**, 286–290 (1989)
5. David, T.E., Omran, A., Armstrong, S., Sun, Z., Ivanov, J.: Long-term results of mitral valve repair for myxomatous disease with and without chordal replacement with expanded polytetrafluoroethylene sutures. J. Thorac. Cardiovasc. Surg. **115**, 1279–1285 (1998). [discussion 85–6]
6. Salvador, L., Mirone, S., Bianchini, R., Regesta, T., Patelli, F., Minniti, G., et al.: A 20-year experience with mitral valve repair with artificial chordae in 608 patients. J. Thorac. Cardiovasc. Surg. **135**, 1280–1287 (2008)
7. von Oppell, U.O., Mohr, F.W.: Chordal replacement for both minimally invasive and conventional mitral valve surgery using premeasured Gore-Tex loops. Ann. Thorac. Surg. **70**, 2166–2168 (2000)

8. Kudo, M., Yozu, R., Kokaji, K., Iwanaga, S.: Feasibility of mitral valve repair using the loop technique. Ann. Thorac. Cardiovasc. Surg. **13**, 21–26 (2007)
9. Votta, E., Le, T.B., Stevanella, M., Fusini, L., Caiani, E.G., Redaelli, A., et al.: Toward patient-specific simulations of cardiac valves: state-of-the-art and future directions. J. Biomech. **46**, 217–228 (2013)
10. Kunzelman, K., Reimink, M.S., Verrier, E.D., Cochran, R.P.: Replacement of mitral valve posterior chordae tendineae with expanded polytetrafluoroethylene suture: a finite element study. J. Card. Surg. **11**, 136–145 (1996)
11. Rim, Y., Laing, S.T., McPherson, D.D., Kim, H.: Mitral valve repair using ePTFE sutures for ruptured mitral chordae tendineae: a computational simulation study. Ann. Biomed. Eng. **42**, 139–148 (2014)
12. Stevanella, M., Maffessanti, F., Conti, C., Votta, E., Arnoldi, A., Lombardi, M., et al.: Mitral valve patient-specific finite element modeling from cardiac MRI: application to an annuloplasty procedure. cardiovasc. Eng. Technol. **2**, 66–76 (2011)
13. Degandt, A.A., Weber, P.A., Saber, H.A., Duran, C.M.: Mitral valve basal chordae: comparative anatomy and terminology. Ann. Thorac. Surg. **84**, 1250–1255 (2007)
14. Lam, J.H., Ranganathan, N., Wigle, E.D., Silver, M.D.: Morphology of the human mitral valve I Chordae tendineae: a new classification. Circulation **41**, 449–458 (1970)
15. May-Newman, K., Yin, F.C.: A constitutive law for mitral valve tissue. J. Biomech. Eng. **120**, 38–47 (1998)
16. Kunzelman, K.S., Einstein, D.R., Cochran, R.P.: Fluid-structure interaction models of the mitral valve: function in normal and pathological states. Philos. Trans. R. Soc. Lond. B. Biol. Sci. **362**, 1393–1406 (2007)
17. Kunzelman, K.S., Cochran, R.P.: Mechanical properties of basal and marginal mitral valve chordae tendineae. ASAIO. Trans. **36**, M405–408 (1990)
18. Votta, E., Maisano, F., Bolling, S.F., Alfieri, O., Montevecchi, F.M., Redaelli, A.: The Geoform disease-specific annuloplasty system: a finite element study. Ann. Thorac. Surg. **84**, 92–101 (2007)
19. Dang, M.C., Thacker, J.G., Hwang, J.C., Rodeheaver, G.T., Melton, S.M., Edlich, R.F.: Some biomechanical considerations of polytetrafluoroethylene sutures. Arch. Surg. **125**, 647–650 (1990)
20. Tesler, U.F., Cerin, G., Novelli, E., Popa, A., Diena, M.: Evolution of surgical techniques for mitral valve repair. Tex. Heart. Inst. J. **36**, 438–440 (2009)
21. Falk, V., Seeburger, J., Czesla, M., Borger, M.A., Willige, J., Kuntze, T., et al.: How does the use of polytetrafluoroethylene neochordae for posterior mitral valve prolapse (loop technique) compare with leaflet resection? A prospective randomized trial. J. Thorac. Cardiovasc. Surg. **136**, 1205 (2008). [discussion-6]
22. Paterson-Brown, S., Cheslyn-Curtis, S., Biglin, J., Dye, J., Easmon, C.S., Dudley, H.A.: Suture materials in contaminated wounds: a detailed comparison of a new suture with those currently in use. Br. J. Surg. **74**, 734–735 (1987)
23. Calafiore, A.M.: Choice of artificial chordae length according to echocardiographic criteria. Ann. Thorac. Surg. **81**, 375–377 (2006)
24. Sarsam, M.A.: Simplified technique for determining the length of artificial chordae in mitral valve repair. Ann. Thorac. Surg. **73**, 1659–1660 (2002)
25. Scorsin, M., Al-Attar, N., Lessana, A.: A novel technique of utilizing artificial chordae for repair of mitral valve prolapse. J. Thorac. Cardiovasc. Surg. **134**, 1072–1073 (2007)
26. Maselli, D., De Paulis, R., Weltert, L., Salica, A., Scaffa, R., Bellisario, A., et al.: A new method for artificial chordae length "tuning" in mitral valve repair: preliminary experience. J. Thorac. Cardiovasc. Surg. **134**, 454–459 (2007)
27. Mandegar, M.H., Yousefnia, M.A., Roshanali, F.: Preoperative determination of artificial chordae length. Ann. Thorac. Surg. **84**, 680–682 (2007)

An Extended Computational Framework to Study Arterial Vasomotion and Its Links to Vascular Disease

Etienne Boileau, Dimitris Parthimos and Perumal Nithiarasu

Abstract A mathematical model of vasomotion is presented in the context of an extended computational framework, to bring new insights into the mechanisms involved in the regulation of vascular tone and arterial function. The approach is based on a number of previously published results, and provides a starting point to a unified method to modelling the pathways to endothelial dysfunction. Results are presented for different scenarios, involving a population of coupled smooth muscle cells on an image-based computational domain.

Keywords Model of vasomotion · Smooth muscle cell (SMC) · Endothelium · Calcium ion Ca^{2+} · Nonlinear complex dynamical system

1 Introduction

Flow affects the blood vessel responses in many ways, through mechanotransduction mechanisms, or by modifying the interactions between blood-borne agonists and cell receptors, especially in the surface boundary layer. In addition to the effects of hæmodynamics and transport, complex interactions exist between the arterial wall components, which are involved in many aspects of vascular adaptation. A better understanding of the principles governing blood-wall interactions and vascular dynamics may prove useful in elucidating some of the mechanochemical aspects of arterial disease. Mathematical description of the ion transport systems that participate

E. Boileau (✉) · P. Nithiarasu
Biomedical Engineering and Rheology Group, Zienkiewicz Centre for Computational Engineering, College of Engineering, Swansea University, Swansea SA2 8PP, UK
e-mail: E.Boileau@swansea.ac.uk

P. Nithiarasu
e-mail: P.Nithiarasu@swansea.ac.uk

D. Parthimos
Cardiff University School of Medicine, Institute of Molecular and Experimental Medicine, Wales Heart Research Institute, Cardiff CF14 4XN, UK
e-mail: Parthimos@cardiff.ac.uk

T. Lenarz and P. Wriggers (eds.), *Biomedical Technology*, Lecture Notes in Applied and Computational Mechanics 74, DOI 10.1007/978-3-319-10981-7_9

in the regulation of vascular tone, and various modelling strategies that incorporate different components of the arterial wall are presented together with computational results.

1.1 Arterial Structure

The arterial wall consists of three tunicæ, or layers: the intima, the media and the adventitia, which are, respectively, an inner, middle and outer sheath common to most blood vessels. The intima is made up of a single layer of endothelial cells (ECs), the endothelium, supported by an internal elastic lamina that separates the intima from the media. Described as an active metabolic and endocrine organ, the endothelium forms a selective permeable membrane, mediating many aspects of blood-tissue exchange. The media lies between the internal and the external elastic laminæ, and is composed of smooth muscle cells (SMCs), embedded in a matrix of elastin and collagen fibres, the amount of elastic tissue varying depending on the size of the vessel. In arterioles, it consists largely of smooth muscle (SM) arranged in lamellæ, wrapped around the vessel. Changes in their contractile tension cause the vessel to dilate or constrict, thereby regulating vessel diameter and blood perfusion. Myoendothelial gap junctions provide a communication pathway between the intima and the media. The adventitia, or tunica externa, is a layer of connective tissue, that often contains sympathetic fibre terminals and, in large arteries and veins, the vasa vasorum.

1.2 Arterial Function, Transport and Wall Dynamics

Vascular health is characterised by the ability of a blood vessel to adapt to the variable requirements of its local environment. Atherosclerosis, coronary and carotid artery disease are all associated with a progressive impairment of this vascular reactivity, or endothelial dysfunction [1–3], a consequence of which may be the long term dysregulation of the mechanisms within the SM associated with vascular tone. Vascular tone, i.e. the tension exerted by the vascular SM, is an essential prerequisite for vasodilation. In skeletal muscle, for example, blood flow can be increased up to 20-fold by vasodilation, to meet local demand of exercise [4]. Vascular arteriolar tone also acts at a local level to regulate capillary recruitment and pressure, enhancing perfusion, nutrient and water exchange. At the systemic level, arterial and central venous pressures are modulated by the continuous adaptation of resistance vessel and peripheral vein tone, respectively. In concert with extrinsic mechanisms, vascular tone is controlled by endothelial factors, vasoactive metabolites and extracellular autocrine and/or paracrine signalling molecules. Vascular blood flow regulation involves a hierarchy of control processes. The endothelium exerts its influence primarily at a local or middle level, by modulating the myogenic response. It acts as an active interface, that translates and amplifies the electrochemical signalling between the lumen-side and the SM, where the arterial wall contractile apparatus is located.

1.2.1 Vasomotion

The vascular SM is responsible for spontaneous fluctuations in vessel calibre that are not attributable to heart rate and blood pressure, and are referred to as 'vasomotion'. Vasomotion was first described more than a century ago, but its physiological significance remains the subject of ongoing debate. The nonlinear nature of the contractile dynamics associated with vasomotion is thought to confer specific hæmodynamic advantage over regular or steady-state flow [5]. At the simplest level, it can be interpreted as a mechanism that continuously redistributes flow to tissue, in order to maintain adequate perfusion. More subtle effects, such as the enhancement of lymphatic drainage and microcirculatory mass transport, may reflect the critical role of smaller arteries and arterioles in regulating interstitial tissue pressure, delivery of oxygen and nutrients and washout of metabolites [6]. Vasomotion becomes particularly prominent in pathological states such as hypoxia and hæmorrhagic shock and may then represent an adaptive dynamic response that maintains or re-establishes flow [7]. Oscillations in the diameter of larger arteries have also been observed in vitro, such as human coronary [8] and pial arteries [9], rabbit small ear [10] and mesenteric arteries [11] or carotid arteries from rats and dogs [12, 13], suggesting a similar origin as that seen in the microcirculation. The interested reader is also referred to the excellent review by Nilsson and Aalkjaer [14].

Vasomotion represents an emergent behaviour associated with a complex dynamical system, whereby coordination of calcium ion Ca^{2+} oscillations in individual SMCs give rise to 'higher-level' phenomena. In isolated vessels, vasomotion exhibits specific patterns of behaviour that are generic to nonlinear physico-chemical systems [15–18]. An important characteristic of nonlinear systems is the co-existence of multiple operating modes with the potential for transitions between them, controlled by single system parameters. State responses of such systems often exhibit completely unpredictable behaviour in response to minute perturbations, allowing the selection of patterns that may confer biological advantage with minimal expenditure of energy.

1.2.2 Characterisation of the Vascular Smooth Muscle

From a dynamical point of view, SMCs are regulators of Ca^{2+}, which is the catalyst of the cells' contractile machinery. Intracellular Ca^{2+} regulation is predominantly controlled by endothelium-mediated polarization and the operating point of ionic transport mechanisms across the cell membrane. For vasomotion to occur, an oscillator must be present; and in order to get macroscopic oscillations of a blood vessel, SMCs' individual oscillations must be synchronized. Although it is well-established that synchronization of Ca^{2+} oscillations in SMCs depends on gap junctions, it remains unclear whether the principal coupling mechanism is electrical or chemical in nature [19].

Background to the Model

The excitation-contraction coupling, or electrochemical coupling, is primarily controlled by movements of Ca^{2+} that permeates from the extracellular space into the cytoplasm. Cytosolic Ca^{2+} concentration ($[Ca^{2+}]_i$) is normally maintained at a basal level of ~100 nm by the plasma membrane Ca^{2+} –ATPase (PMCA), the $Na^{+}-Ca^{2+}$ exchanger (NCX) and the sarco endoplasmic reticulum Ca^{2+} –ATPase (SERCA). Vasoconstricting stimuli initiate SM contraction by increasing the $[Ca^{2+}]_i$, to reach the μM range. Extensive pharmacological probing has characterised vasomotion as the interplay between a 'slow, intracellular' Ca^{2+} oscillator (period ~1–5 min) and a 'fast, membrane' oscillator (period 5–30 s) operating within the SMC layer of the arterial wall [15, 20]. The intracellular oscillator is identified by the cyclic Ca^{2+}-induced Ca^{2+} release (CICR) from ryanodine-sensitive stores of the sarco endoplasmic reticulum (SR/ER). Morphologically, the SR/ER of the SM is spatially heterogeneous and possesses both Ca^{2+}- and inositol 1,4,5-trisphosphate ($InsP_3$)-sensitive Ca^{2+} stores. In addition to the action of CICR and $InsP_3$-induced Ca^{2+} release (IICR), Ca^{2+} release from the stores is also due to a passive leak from the SR. Refilling of the stores is achieved by the SERCA pump. SERCA pump blockers, such as cyclopiazonic acid and thapsigargin, deplete the Ca^{2+} stores and eliminate the slow/large amplitude component of vasomotion [17, 20]. After each discharge the stores are replenished by Ca^{2+} influx across the cell membrane. These Ca^{2+} currents are primarily conducted via voltage-operated channels (VOCCs), nonspecific cation channels (NSCCs) and Na^{+}-Ca^{2+} exchange via the NCX. Ca^{2+} released by the SR is extruded from the cytosol via the membrane Ca^{2+} extrusion ATPase, and inhibition of the pump with vanadate produces large constrictor response due to increased $[Ca^{2+}]_i$ [16]. Ca^{2+} extrusion can also be performed by the NCX with a stoichiometry of 3 $Na^{+}$for 1 Ca^{2+} [15, 21]. The exchanger can act both as an efflux or an influx (forward/reverse mode respectively) based on the membrane potential being above or below the reversal potential of NCX.

Membrane potential is a distinct dynamic variable determined by the additive contribution of a large number of ionic transport mechanisms (ion channels, pumps, and exchangers). Suppression of Ca^{2+}-activated $K^{+}(K_{Ca})$ channels (with tetraethylammonium and charybdotoxin), Cl^{-}channels (by reducing extracellular Cl^{-}concentration and niflumic acid), the $Na^{+}-K^{+}$–ATPase (by ouabain), or Na^{+}/Ca^{2+} exchange (by low extracellular Na^{+}) consistently attenuate the fast membrane component of arterial vasomotion, suggesting that these transport mechanisms are fundamental in the genesis of vasomotion [22]. The fundamental ionic transport mechanisms involved in the homeodynamics of intracellular Ca^{2+} are presented in Fig. 1.

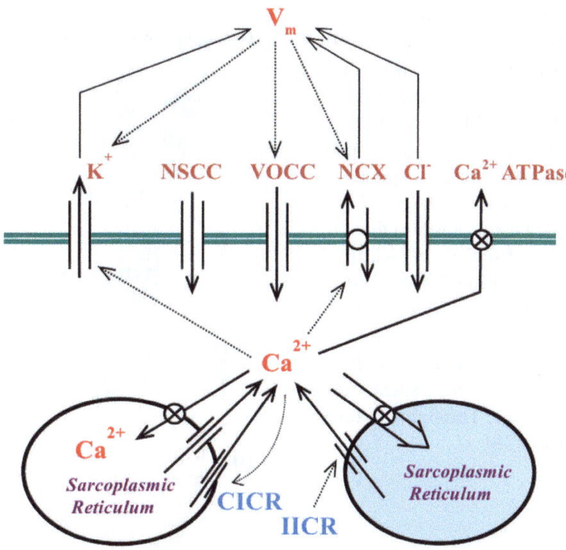

Fig. 1 Schematic of a SMC illustrating the coupling between the store-mediated intracellular Ca^{2+} oscillator and membrane potential fluctuations due to ionic transport mechanisms across the cellular membrane. The system of ordinary differential equations describing these processes is presented below

1.2.3 The Role of the Endothelium

Through their direct and continuous contact to blood flow, ECs are exposed to the action and combined effect of both mechanical forces and biochemical stimulation. Electrical and chemical signalling between endothelial and SM cells contribute to arterial function via two primary interrelated mechanisms. Longitudinal propagation of dilations-constrictions along the vessel wall constitutes one of these, while the other comprises the effects of the endothelium-derived hyperpolarization factor (EDHF), in which transmission of endothelial hyperpolarization to the SM mediates relaxation. Both pathways are dependent on direct intercellular coupling via gap junctions, allowing the diffusion of ions and small molecules (1 kDa in size) and conferring electrical continuity between coupled cells [23, 24]. The role of the endothelium in vasomotion nevertheless remains unclear. Some studies reported rhythmic contractions in the absence of an intact endothelium, whereas others claimed that its presence is needed for synchronization [25–28]. Such apparently paradoxical findings may reflect the unpredictability that is characteristic of nonlinear systems, and can potentially be clarified by further modelling studies.

Flow Regulation of Agonist-Ca^{2+} Coupling

Although the results presented below do not include any coupling between endothelial and SM cells, future work will address this point, in addition to considering the effects of mass transport on mobilization of endothelial $[Ca^{2+}]_i$. Using adenine

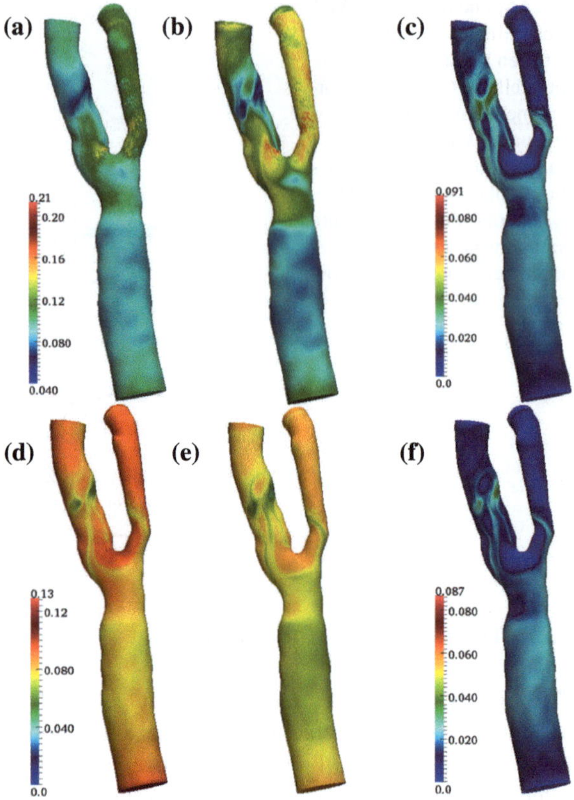

Fig. 2 *Upper row* Distribution on the arterial surface of [ATP] (μM) during **a** systole **b** diastole and [ADP] (μM) during **c** diastole. ATP production at the endothelium depends on wall shear stress only. *Lower row* Distribution on the arterial surface of [ATP] (μM) during **d** systole, **e** diastole and [ADP] (μM) during **f** diastole. ATP production at the endothelium depends on wall shear stress and time. The figure is taken from [38]

nucleotides as the agonist (mostly adenosine-5' triphosphate ATP and adenosine diphosphate ADP), many researchers have tried to characterise flow regulation of ATP- and ADP-Ca^{2+} coupling in the concentration boundary-layer, and the effect of enzymatic degradation at the endothelium [29–37]. Small changes in the model parameters, such as the rate constants for convection of ATP from the bulk fluid to the cell surface or for the degradation and/or production of ATP significantly influence the pattern of catalytic reactions at the endothelium [38]. As we have shown earlier, and as illustrated in Fig. 2, even though the bulk concentration of secreted agonists is low and mostly unaffected by the flow, variations exist at the endothelium, which are nevertheless difficult to quantify numerically. Higher concentration of agonists, and the existence of steep gradients in the boundary layer will be affected by changes in the flow field, potentially generating further signalling cascades. Such propagated effects might constitute elements of an intravascular axis that serves to regulate vascular tone [38–42]. See also the review by Davies on endothelial mechanotransduction [43].

1.2.4 Link to Cardiovascular Disease

We previously demonstrated that specific dynamical transitions, associated phenomenologically with cardiovascular disease, can be readily effected by pharmacological manipulation of cellular mechanisms (e.g. irregular Ca^{2+} inflow or outflow from the SR/ER via the SERCA and RyR pumps respectively) [44]. Such mechanisms can be probed in a rigorous fashion, providing a broad range of relevant experimental data for model development and validation. There is substantial evidence showing that the prevalence of vasomotion can indeed be altered under specific pathological conditions. In a large proportion of diabetic patients, for example, vasomotion is impaired, and the amplitude of flow-motion is also reduced [45, 46].

The formation of atheromatic plaque in stenosed vessels is another central area of modelling investigation [47–49]. In addition to the effects of hæmodynamics and transport, in early atheroma, there exists complex interactions between the arterial wall components (SMCs and ECs), which cause inflammatory signalling that leads to monocyte accumulation, foam cell degeneration and formation of atheromatous plaque. Such alterations can result in significant modification of the arterial geometry. These contributory factors can thus be incorporated in order to gain insights into potential causes of vascular adaptation.

The role of mathematical/computational modelling in studying cardiovascular disease as an integrated systemic dysfunction is therefore very important. Comprehensive studies of the mechanisms involved in the genesis of disease will require concurrent modelling of endothelial factors, electrical coupling, calcium synchronization and wave propagation as critical components. A first attempt in this direction is presented here.

2 Model

The intracellular oscillator (outlined previously) is modelled as CICR from the SR via ryanodine sensitive Ca^{2+} channels. This formulation is based on experimental evidence that CICR/IICR can be attributed to distinct store subtypes in rabbit ear arteries [17, 20]. The ryanodine-sensitive channel is modelled as a product of two sigmoidal functions that account for $[Ca^{2+}]$ in the cytosol and $[Ca^{2+}]$ in the SR, in agreement with experimental evidence [50]. There are contradictory observations that Ca^{2+} oscillations may occur under constant elevated levels of $InsP_3$, or under fluctuating concentrations of intracellular $InsP_3$. Considering the dominant modulatory effect of ryanodine on arterial dynamics observed in rabbit ear arteries, we assume Ca^{2+} release from $InsP_3$-sensitive stores to be a constant that increases directly with concentrations of histamine. The effect of IICR is thus included in the coefficient A, that corresponds to Ca^{2+} uptake/release in the cytosol. Sequestration of Ca^{2+} by the SERCA pump is represented by a sigmoidal function of intracellular $[Ca^{2+}]$. A passive leak from the SR is included to account for experimental evidence, although it has limited effect on the oscillatory dynamics under physiological conditions [51]. The dominant cell Ca^{2+} influx pathway in arterial SM is provided by VOCCs. Voltage sensitivity of Ca^{2+} refilling can account for large-scale vascular contractile activity and synchronization, under conditions of electrical cell-cell coupling [19]. With a reversal potential of ~0.1 V and a sigmoidal open state probability centred at ~−0.024 V, VOCCs are maximally active during cellular depolarization, a state generally associated with enhanced vascular contraction [52]. In spite of a central role in Ca^{2+} transport and the generation of contractile activity, VOCCs account for only up to 10 % of cellular polarization. A number of ionic transport mechanisms that are fundamental in polarization of the arterial wall have also been included in the quantification of SM membrane potential. These are the Cl^-channel, the NCX, known to be central in cardiac muscle contraction, and the K^+channels. With a reversal potential of ~−0.095 V, K^+channels are particularly important in vascular dynamics since they constitute the main hyperpolarizing force in the arterial wall [22]. A large number of Ca^{2+}-activated K^+channel subtypes have been identified, conventionally grouped as large, intermediate and small conductance channels. In the present formulation, we have grouped all K^+transport activity under a single idealised channel subtype with a linear voltage dependence and a sigmoidal Ca^{2+} activation. Typically, ionic transport mechanisms are both Ca^{2+}- and voltage-sensitive, as indeed reflected in the mathematical formulation of each transport component.

2.1 Coupled Intracellular and Membrane Ca^{2+} Oscillators in Smooth Muscle Cells

Let $x = [Ca^{2+}]_i$ represent the cytosolic free Ca^{2+} concentration, $y = [Ca^{2+}]_{SR}$ the Ca^{2+} concentration in the sarcoplasmic reticulum and z the cell membrane potential. The system described above and depicted in Fig. 1 is represented by the following equations:

$$\begin{aligned}\frac{dx}{dt} = & \underbrace{A}_{\substack{\text{NSCC}\\ \text{influx}}} - \underbrace{E_{Ca}\frac{z - z_{Ca1}}{1 + e^{-(z-z_{Ca2})/R_{Ca}}}}_{\text{VOCC influx}} + \underbrace{E_{Na/Ca}\frac{x}{x + x_{Na/Ca}}(z - z_{Na/Ca})}_{\text{NCX}} \\ & - \underbrace{B\frac{x^n}{x^n + x_b^n}}_{\text{SR uptake}} + \underbrace{C_r\frac{x^{p_r}}{x^{p_r} + x_r^{p_r}}\frac{y^{m_r}}{y^{m_r} + y_r^{m_r}}}_{\text{RyR CICR}} \\ & - \underbrace{Dx^k\left(1 + \frac{z - z_d}{R_d}\right)}_{Ca^{2+}\text{ extrusion}} + \underbrace{Ly}_{\text{SR leak}} \end{aligned} \tag{1a}$$

$$\frac{dy}{dt} = \underbrace{B\frac{x^n}{x^n + x_b^n}}_{\text{SR uptake}} - \underbrace{C_r\frac{x^{p_r}}{x^{p_r} + x_r^{p_r}}\frac{y^{m_r}}{y^{m_r} + y_r^{m_r}}}_{\text{RyR CICR}} \tag{1b}$$

$$\frac{dz}{dt} = -\gamma \left(\underbrace{E_{Cl}\frac{x}{x+x_{Cl}}(z-z_{Cl})}_{Cl^{-}\text{channels}} + \underbrace{2E_{Ca}\frac{z-z_{Ca1}}{1+e^{-(z-z_{Ca2})/R_{Ca}}}}_{\text{VOCC influx}} \right.$$
$$\left. + \underbrace{E_{Na/Ca}\frac{x}{x+x_{Na/Ca}}\left(z-z_{Na/Ca}\right)}_{\text{NCx}} + \underbrace{E_K\left(z-z_K\right)\frac{x}{x+\beta e^{-(z-z_{Ca3})/R_K}}}_{K^{+}\text{efflux}} \right) \quad (1c)$$

where the electric reversal potentials with respect to Ca^{2+} and $Na^{+}$are determined from the Nernst equation according to:

$$z_{Ca1} = \frac{RT}{FZ_{Ca}} ln \left(\frac{[Ca^{2+}]_{i,0}}{x} \right) \qquad z_{Na} = \frac{RT}{FZ_{Na}} ln \left(\frac{[Na^{+}]_{i,0}}{[Na^{+}]_i} \right) \quad (2)$$

and where Z_{Ca}=1, $Z_{Na} = 2$, RTF^{-1}=0.024, $(RT/FZ_{Na})\, ln\left([Na^{+}]_{i,0}/[Na^{+}]_i\right) =$ 0.069 V, $[Ca^{2+}]_{i,0} = 0.0025\mu M$, R is the gas constant in J K^{-1} mol^{-1}, F is the Faraday constant in A s mol^{-1} and $T = 310$ K. The reversal potential of the NCX is calculated as $z_{NCx} = 3z_{Na} - 2z_{Ca1}$. All values of the coefficients are found in Table 1. A complete description of the mathematical model can be found in [19, 51]. The subscript r refers to RyR-mediated CICR, as opposed to IICR (not included in the present formulation). The same general conclusions would not significantly differ whether the CICR positive feedback mechanism is modelled as an InsP$_3$- or RyR-dependent process, as was also pointed out in [19]. Admittedly, the addition of InsP$_3$ as another independent variable could allow to asses the importance of crosstalks between RyR and Insp$_3$R, and extend the present formulation to include coupling between the endothelium-fluid domain and the SM layer.

2.2 Cell-Cell Coupling

The model has been extensively validated, with analysis and results presented for the case of a single SMC [19, 51]. The generalization of these results to a population of SMCs, using image-based geometries, allows to characterize the effects of a number of parameters, while enabling a more 'intuitive representation' of the phenomenon.

The nature of coupling between adjacent cells remains controversial. In the current formulation we explore the consequences of either Ca^{2+} or electrical coupling, by assuming a simple gradient driven flux between neighbouring cells coupled via gap junctions. Two terms:

$$J_{Ca,i} = g_{Ca} \sum_{j \in N_i} \left(x_j - x_i\right) \qquad V_{m,i} = g_z \sum_{j \in N_i} \left(z_j - z_i\right) \quad (3)$$

are added to (1a) and (1c), respectively, and for each SMC i, to model Ca^{2+} and electrical coupling with all nearest neighbouring SMCs $j \in N_i$. Depending on

Table 1 Coefficients

Parameter	Description	Value
A	Ca^{2+} influx via NSCC	$2.3\,\mu M\,s^{-1}$
L	SR leak rate constant	$0.025\,s^{-1}$
γ	scaling factor (inversely related to cell capacitance)	$1\,V\,\mu M^{-1}$
VOCC influx		
E_{Ca}	whole cell conductance	$12\,\mu M\,V^{-1}\,s^{-1}$
z_{Ca1}	reversal potential	0.12 to 0.135 V
z_{Ca2}	half point of activation sigmoid	−0.024 V
R_{Ca}	max. slope of activation sigmoid	0.0085 V
NCX		
$E_{Na/Ca}$	whole cell conductance	$43.8\,\mu M\,V^{-1}\,s^{-1}$
$z_{Na/Ca}$	reversal potential	−0.03 to −0.045 V
$x_{Na/Ca}$	half point of Ca^{2+} activation	$0.5\,\mu M$
SR uptake		
B	rate constant	$400\,\mu M\,s^{-1}$
x_b	half point of ATPase activation sigmoid	$4.4\,\mu M$
n	Hill coefficient	2
RyR CICR		
C_r	rate constant	$1250\,\mu M\,s^{-1}$
y_r	half point of Ca^{2+} efflux sigmoid	$8.9\,\mu M$
x_r	half point of CICR activation sigmoid	$0.9\,\mu M$
m_r	Hill coefficient	2
p_r	Hill coefficient	4
Ca^{2+} extrusion by ATPase pump		
D	rate constant	$6.25\mu M\,s^{-1}$
z_d	intercept of voltage dependence	−0.1 V
R_d	slope of voltage dependence	0.25 V
k	exponent for $[Ca^{2+}]_i$ dependence	2
Cl^- channels		
E_{Cl}	whole cell conductance	$65\,\mu M\,V^{-1}\,s^{-1}$
z_{Cl}	reversal potential	−0.025 V
x_{Cl}	Ca^{2+} sensitivity	$0\,\mu M$
k	exponent for $[Ca^{2+}]_i$ dependence	2
K^+ efflux		
E_K	whole cell conductance	$43\,\mu M\,V^{-1}\,s^{-1}$
z_K	reversal potential	−0.095 V
z_{Ca3}	half point of K_{Ca} channel activation sigmoid	−0.027 V
R_K	max. slope of K_{Ca} channel activation sigmoid	0.012 V
β	Ca^{2+} sensitivity of K_{Ca} channel activation sigmoid	$0\,\mu M$

the surface discretization, or on the modelling strategy, N_i may represent the set of connected surface points belonging to a patch centred on point (cell) i. The geometry used for the simulation is a carotid artery, and has a total surface area of $8.26\,\text{cm}^2$, giving a mean cell surface area of $\sim 143\,\mu^2$. The actual geometry of each SMC is not modelled explicitly in the present formulation.

2.2.1 Ca^{2+} Coupling

All cells are dynamically identical except for the A parameter, which reflects influx of Ca^{2+} via NSCCs, and is randomly distributed around its mean value, given in Table 1. Parameter A was chosen for its strong modulatory effect on the oscillatory frequency. A typical cellular matrix thus contains SMCs that are quiescent (either in an under-stimulated or an over-stimulated state, in terms of Ca^{2+} availability) and cells that are oscillating spontaneously. Within this range, the oscillatory regime is maintained at a physiological level.

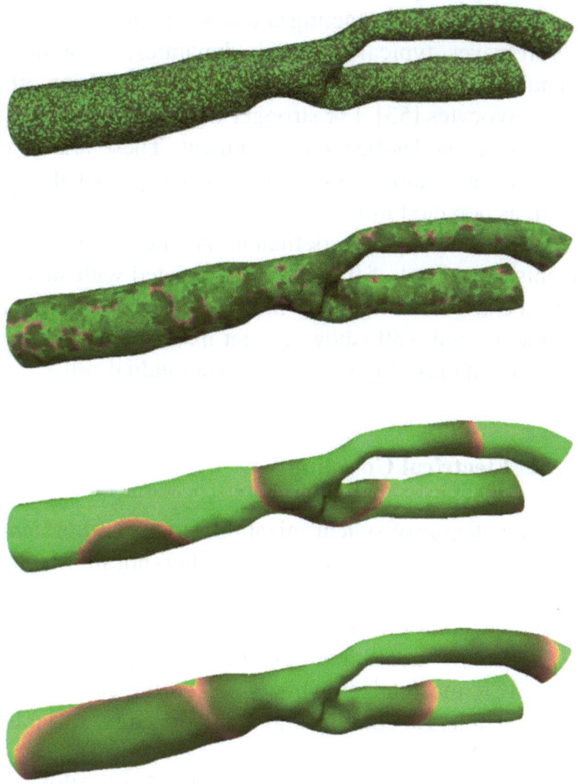

Fig. 3 Computational modelling results illustrating the emergence of propagating wave fronts. A gradual transition of uncorrelated oscillatory activity into coherent patterns is observed under conditions of increased intercellular Ca^{2+} coupling (From *top* to *bottom* $g_{Ca} = 0$, $g_{Ca} = 2$, $g_{Ca} = 10$, $g_{Ca} = 20$). Colours are coded as: red = cytosolic free $[Ca^{2+}]_i$, green = $[Ca^{2+}]_{SR}$, blue = membrane potential. A magenta colour is seen where red and blue strengths are equal, and therefore indicative of membrane depolarization and high Ca^{2+} influx

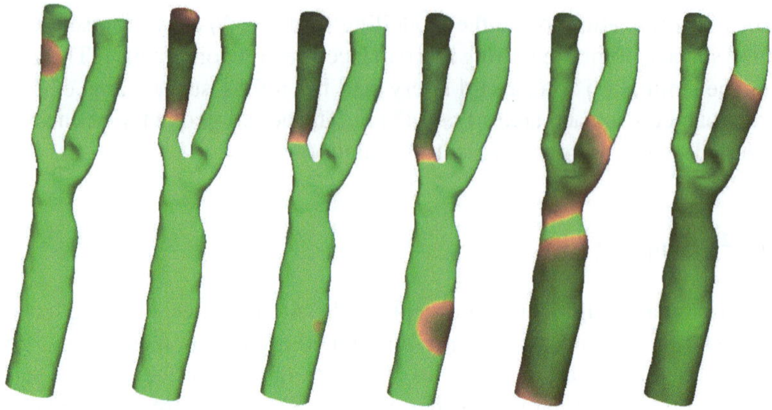

Fig. 4 Computational modelling results illustrating the emergence of propagating wave fronts for a stronger Ca^{2+} coupling (g_{Ca} = 50). Colour coded as in Fig. 3

Weakly coupled cells ($g_{Ca} = 2$) are dominated by local dynamics, producing a complex field of interacting and non-repetitive waveforms characterised by spiral and scroll waves, typical of chaotic dynamics. These highly complex, yet deterministic in nature, types of oscillatory behaviour have been often observed in syncytia of cardiac myocytes [53]. For stronger Ca^{2+} coupling ($g_{Ca} = 10$ and $g_{Ca} = 20$), discrete pacemaker nodes become prominent. These are eventually dominated by a single pacemaker resulting in clear wave fronts propagated longitudinally along the arterial wall, as depicted in Fig. 3. A propagating wave is illustrated in Fig. 4, for a stronger value of the coupling coefficient. As the oscillatory behaviour emerges, ion movements in and out of the SR are associated with successive high and low $[Ca^{2+}]_{SR}$. This behaviour coincides with the synchronization of intracellular Ca^{2+} oscillations of the arterial wall, showing that macroscopic rhythmic activity can indeed reflect ion movements at the level of the individual cell.

2.2.2 Electrical Coupling

Limited degree of synchronization, mostly at short length scales, was observed under conditions of purely electrical coupling (not shown). This is due to the present formulation of the model, which is expressing primarily the dynamics of intracellular CICR. The inclusion of a specific ionic transport process as a further independent variable would allow membrane potential to support oscillatory activity independently of the CICR mechanism. This is indeed the case when the open state probability of K_{Ca} channels is employed as an independent variable in a four differential equation model [51].

The interplay between extracellular Ca^{2+} influx via NSCCs and VOCCs was also investigated. Independently of the coupling parameters g, as given in (3), simulations showed how low or high influx through either channel leads to the dominance of Ca^{2+} and/or electrical coupling, and to the loss of oscillatory activity. For the case of a single cell model, the effects of variations in the control parameters A and E_{Ca} were investigated by nullcline analysis in [19].

3 Integrated Approach

The modelling framework presented here can be used in realistic scenarios, to gain insights into the fundamental physiology of inter-cellular coupling, by analysing the emergence of synchronization and macroscopic wave propagation. The theoretical consequences of local neighbour interactions through electrical and/or chemical coupling via inter-cellular diffusion of Ca^{2+} ions have only been briefly explored, and can be further compared using the Nernst-Planck equation. This approach may also identify key dynamical features, by exploring stability characteristics and critical scaling phenomena near the onset of synchronization, as described for Kuramoto-type phase synchronization of oscillators [54], and the mode locking observed in nonlinear systems [55].

Additional components will include the coupling between the SM and the endothelium, especially with a view to provide a robust basis for the study of various patho-physiological scenarios. Further manifestations of blood-wall and arterial dynamics coupling could then be investigated, by manipulation of a number of potential mechanisms underlying flow-endothelium interactions. Such mechanisms may include wall shear stress, which can modulate calcium responses in SMCs indirectly, via mass transfer of agonist and receptor-ligand interactions, as described in Sect. 1.2.3, and directly through wall shear stress operated channels. In this context, ATP and its derived products can be considered. Potential mechanisms of action may include: (i) ATP as a mediator of Ca^{2+} extrusion ATPase, (ii) the sodium-potassium exchanger Na^+/K^+-ATPase involved in maintaining the cell membrane potential, and (iii) ATP-sensitive potassium (K_{ATP}) channels in the endothelium. Such an integrated approach will enable the active coupling between blood flow and arterial wall dynamics, and provide a uniquely global view of cardiovascular physiology that has not been achieved before in computational fluid flow studies.

References

1. Halcox, J.P.J., Donald, A.E., Ellins, E.: Endothelial function predicts progression of carotid intima-media thickness. Circ. **119**, 1005–1012 (2009)
2. Halcox, J.P.J., Schenke, W.H., Zalos, G.: Prognostic value of coronary vascular endothelial dysfunction. Circ. **106**, 653–658 (2002)
3. Schachinger, V., Britten, M.B., Zeiher, A.M.: Prognostic impact of coronary vasodilator dysfunction on adverse long-term outcome of coronary heart disease. Circ. **101**, 1899–1906 (2000)

4. Levick, J.R.: An Introduction to Cardiovascular Physiology, 5th edn. Hodder Arnold, London (2010)
5. Parthimos, D., Edwards, D.H., Griffith, T.M.: Comparison of chaotic and sinusoidal vasomotion in the regulation of microvascular flow. Cardiovasc. Res. **31**, 388–399 (1996)
6. Griffith, T.M.: Temporal chaos in the microcirculation. Cardiovasc. Res. **31**, 342–358 (1996)
7. Intaglietta, M.: Arteriolar vasomotion: implications for tissue ischæmia. Blood Vessels **28**, 1–7 (1991)
8. Kawasaki, K., Seki, K., Hosoda, S.: Spontaneous rhythmic contractions in isolated human coronary arteries. Experientia **37**, 1291–1292 (1981)
9. Gokina, N.I., Bevan, R.D., Walters, C.L., Bevan, J.A.: Electrical activity underlying rhythmic contraction in human pial arteries. Circ. Res. **78**, 148–153 (1996)
10. Griffith, T.M., Edwards, D.H.: Mechanisms underlying chaotic vasomotion in isolated resistance arteries: roles of calcium and EDRF. Biorheology **30**, 333–347 (1993)
11. Omote, M., Kajimoto, N., Mizusawa, H.: The ionic mechanism of phenylephrine-induced rhythmic contractions in rabbit mesenteric arteries treated with ryanodine. Acta Physiol. Scand. **147**, 9–13 (1993)
12. Masuda, Y., Okui, K., Fukuda, Y.: Fine spontaneous contractions of the arterial wall of the rat in vitro. Jpn. J. Physiol **32**, 453–457 (1982)
13. Siegel, G.: Principles of vascular rhythmogenesis. Progr. Appl. Microcirc. **3**, 40–63 (1983)
14. Nilsson, H., Aalkjaer, C.: Vasomotion: mechanisms and physiological importance. MI **3**, 79–89 (2003)
15. Griffith, T.M., Edwards, D.H.: Fractal analysis of the role of smooth muscle Ca2+ fluxes in genesis of chaotic arterial pressure oscillations. Am. J. Physiol. **266** H1786–H1800 (1994) (Heart Circ. Physiol. 35)
16. Griffith, T.M., Edwards, D.H.: EDRF suppresses chaotic pressure oscillations in isolated resistance artery without influencing intrinsic complexity. Am J. Physiol. **266** H1801–H1811 (1994) (Heart Circ. Physiol. 35)
17. Griffith, T.M., Edwards, D.H.: Ca sequestration as a determinant of chaotic and mixed-mode dynamics in agonist-induced vasomotion. Am J. Physiol. **272** H1696–H1709 (1997) (Heart Circ. Physiol. 41)
18. Griffith, T.M., Edwards, D.H.: Entrained ion transport systems generate the membrane component of chaotic agonist-induced vasomotion. Am J. Physiol. **273** H909–H920 (1997) (Heart Circ. Physiol. 42)
19. Parthimos, D., Haddock, R.E., Hill, C.E., Griffith, T.M.: Dynamics of a three-variable nonlinear model of vasomotion: comparison of theory and experiments. Biophys. J. **93**, 1534–1556 (2007)
20. De Brouwer, S., Edwards, D.H., Griffith, T.M.: Simplification of the quasiperiodic route to chaos in agonist-induced vasomotion by iterative circle maps. Am. J. Physiol. **274**, H1315–H1326 (1998)
21. O'Donnell, M.E., Owen, N.E.: Regulation of ion pumps and carriers in vascular smooth muscle. Physiol. Rev. **74**, 683–721 (1994)
22. Toro, L., Vaca, L., Stefani, E.: Calcium-activated potassium channels from coronary smooth muscle reconstituted in lipid bilayers. Am. J. Physiol. **260**, H1779–H1789 (1991)
23. Griffith, T.M.: Endothelium-dependent smooth muscle hyperpolarization: do gap junctions provide a unifying hypothesis? Pharmacology **141**, 881–903 (2004)
24. Haddock, R.E., Grayson, T.H., Brackenbury, T.D., Meaney, K.R., Neylon, C.B., Sandow, S.L., Hill, C.E.: Endothelial coordination of cerebral vasomotion via myoendothelial gap junctions containing connexins 37 and 40. Am. J. Physiol. **291**, H2047–H2056 (2006)
25. Peng, H., Matchkov, A., Ivarsen, A., et al.: Hypothesis for the initiation of vasomotion. Circ. Res. **88**, 810–815 (2001)
26. Sell, M., Boldt, W., Markwardt, F.: Desynchronising effect of the endothelium on intracellular Ca2+ concentration dynamics in vascular smooth muscle cells of rat mesenteric arteries. Cell Calcium **32**, 105–120 (2002)
27. Okazaki, K., Seki, S., Kanaya, J., et al.: Role fo the endothelium-derived hyperpolarizing factor in phenylephrine-induced oscillatory vasomotion in rat small mesenteric artery. Anesthesiology **98**, 1164–1171 (2003)

28. Rahman, A., Matchkov, V., Nilsson, H., Aalkjaer, C.: Effects of cGMP on coordination of vascular smooth muscle cells of rat mesenteric small arteries. J. Vasc. Res. **42**, 301–311 (2005)
29. Dull, R.O., Davies, P.F.: Flow modulation of agonist (ATP)-response (Ca2+) coupling in vascular endothelial cells. Am. J. Physiol. **261**, 149–154 (1991)
30. Dull, R.O., Tarbell, J.M., Davies, P.F.: Mechanisms of flow-mediated signal transduction in endothelial cells: kinetics of ATP surface concentrations. J. Vasc. Res. **28**, 410–419 (1992)
31. Nollert, M.U., Diamond, S.L., McIntire, L.V.: Hydrodynamic shear stress and mass transport modulation of endothelial cell metabolism. Biotech. Bioeng. **38**, 588–602 (1991)
32. Shen, J., Gimbrone Jr, M.A., Luscinskas, F.W., Dewey Jr, C.F.: Regulation of adenine nucleotide concentration at endothelium-fluid interface by viscous shear flow. Biophys. J. **64**, 1323–1330 (1993)
33. John, K., Barakat, A.I.: Modulation of ATP/ADP concentration at the endothelial surface by shear stress: effect of flow-induced ATP release. Ann. Biomed. Eng. **29**, 740–751 (2001)
34. Choi, H.W., Ferrara, K.W., Barakat, A.I.: Modulation of ATP/ADP concentration at the endothelial surface by shear stress: effect of flow recirculation. Ann. Biomed. Eng. **35**, 505–516 (2007)
35. David, T.: Wall shear stress modulation of ATP/ADP concentration at the endothelium. Ann. Biomed. Eng. **31**, 1231–1237 (2003)
36. Plank, M.J., Wall, D.N.J.: Atherosclerosis and calcium signalling in endothelial cells. Prog. Biophys. Mol. Bio. **91**, 287–313 (2006)
37. Hu, X., Xiang, C., Cao, L.L.: A mathematical model for ATP-mediated calcium dynamics in vascular endothelial cells induced by fluid shear stress. Appl. Math. Mech. **29**, 1291–1298 (2008)
38. Boileau, E., Bevan, R.L.T., Sazonov, I., Nithiarasu, P.: Flow-induced ATP release in patient-specific arterial geometries: a comparative study of computational models. Intl. J. Numer. Method. Biomed. Eng. **29**, 1038–1056 (2013)
39. Buxton, I.L., Kaiser, R.A., Oxhorn, B.C., Cheek, D.J.: Evidence supporting the nucleotide axis hypothesis: ATP release and metabolism by coronary endothelium. Am. J. Physiol. Heart Circ. Physiol. **281**, 1657–1666 (2001)
40. Yang, S., Cheek, D.J., Westfall, D.P., Buxton, I.L.: Purinergic axis in cardiac blood vessels. Agonist-mediated release of ATP from cardiac endothelial cells. Circ. Res. **74**, 401–407 (1994)
41. Duza, T., Sarelius, I.H.: Conducted dilations initiated by purines in arterioles are endothelium dependent and require endothelial calcium. Am. J. Physiol. Heart Circ. Physiol. **285**, 26–37 (2003)
42. Arciero, J.C., Carlson, B.E., Secomb, T.W.: Theoretical model of metabolic blood flow regulation: roles of ATP release by red blood cells and conducted responses. Am. J. Physiol. Heart Circ. Physiol. **295**, 1562–1571 (2008)
43. Davies, P.F.: Flow-mediated endothelial mechanotransduction. Physiol. Rev. **75**, 519–560 (1995)
44. George, C.H., Parthimos, D., Silvester, N.C.: A network-oriented perspective on cardiac calcium signaling. Am. J. Physiol. Cell Physiol. **303**, C897–C910 (2012)
45. Stansberry, K.B., Shapiro, S.A., Hill, M.A., et al.: Impaired peripheral vasomotion in diabetes. Diab. Care **19**, 715–721 (1996)
46. Parthimos, D., Schmiedel, O., Harvey, J.N., Griffith, T.M.: Deterministic nonlinear features of cutaneous perfusion are lost in diabetic subjects with neuropathy. Microvasc. Res. **82**, 42–51 (2011)
47. El Khatib, N., Genieys, V., Volpert, V.: Atherosclerosis initiation modeled as an inflammatory process. Math. Model. Nat. Phenom. **2**, 126–141 (2007)
48. Hoskins, P.R., Hardman, D.: Three-dimensional imaging and computational modelling for estimation of wall stresses in arteries. Br. J. Radiol. **82**, S3–S17 (2009)
49. Cilla, M., Pena, E.: Mathematical modelling of atheroma plaque formation and development in coronary arteries. J. R. Soc. Interface. **11**(90), 20130866 (2014)
50. Mukherjee, S., Thomas, N.L., Williams, A.J.: A mechanistic description of gating of the human cardiac ryanodine receptor in a regulated minimal environment. J. Gen. Physiol. **140**, 139–158 (2012)

51. Parthimos, D., Edwards, D.H., Griffith, T.M.: Minimal model of arterial chaos generated by coupled intracellular and membrane Ca2+ oscillators. Am. J. Physiol. **277**, H1119–H1144 (1999)
52. Nelson, M.T., Patlak, J.B., Worley, J.F., Standen, N.B.: Calcium channels, potassium channels, and voltage dependence of arterial smooth muscle tone. Am. J. Physiol. **259**, C3–C18 (1990)
53. Chang, M.G., Zhang, Y., Chang, C.Y., Xu, L., Emokpae, R., Tung, L., Marbán, E., Abraham, M.R.: Spiral waves and reentry dynamics in an in vitro model of the healed infarct border zone. Circ. Res. **105**, 1062–1071 (2009)
54. Strogatz, S.H.: Exploring complex networks. Nature **410**, 268–276 (2001)
55. Boccaletti, S., Pecora, L.M., Pelaez, A.: Unifying framework for synchronization of coupled dynamical systems. Phys. Rev. E. **63**, 066219 (2001)

Development of a Model of the Electrically Stimulated Cochlea

Waldo Nogueira, Waldemar Würfel, Richard T. Penninger and Andreas Büchner

Abstract Cochlear Implants (CIs) are implantable medical devices that can restore the sense of hearing in people with profound sensorineural hearing loss. Clinical trials assessing speech intelligibility in CI users have found large inter subject variability. One possibility to explain the variability are the individual differences in the interface created between electrodes and the auditory nerve. For example, the exact position of the electrodes in each cochlea may differ from one patient to another. Additionally the amount of functional auditory neurons might also vary considerably between CI users. In order to understand the variability, models of the voltage distribution of the electrically stimulated cochlea may be useful. With this purpose we have developed a model that allows to simulate the voltage distribution at different positions on the auditory nerve. Simulations show differences in the extracellular voltage of the spiral ganglions depending on the electrode positions and the cochlear size, which might explain some of the variability. Finally, the model of the electrically stimulated cochlea has been used to simulate the extracellular voltage patterns produced by different instrumental sounds. These patterns have been inserted in an automatic instrument classifier that helps to illustrate the mentioned variability.

Keywords Cochlear implant · Finite element method · Electric field · Cochlea · Sound coding strategy · Instrument identification

1 Introduction

Cochlear implants (CIs) are medical implantable devices that are used to restore the sense of hearing for people with profound sensorineural hearing loss given that the auditory anatomy is fully developed [21]. Cochlear implantation (and subsequent rehabilitation) typically allows even children with prelingual deafness to develop spoken language understanding and production [4, 15]. In CIs the auditory nerves are

W. Nogueira (✉) · W. Würfel · R.T. Penninger · A. Büchner
Department of Otolaryngology, Medical School Hannover, Cluster of Excellence "Hearing4all", Karl-Wiechert-Allee 3, 30625 Hannover, Germany
e-mail: nogueiravazquez.waldo@mh-hannover.de

T. Lenarz and P. Wriggers (eds.), *Biomedical Technology*, Lecture Notes in Applied and Computational Mechanics 74, DOI 10.1007/978-3-319-10981-7_10

directly stimulated, bypassing the natural functioning of the outer, middle and inner ear. Current is applied to one electrode mimicking the action of the basilar membrane and the inner hair cells. In a natural cochlea, a pure tone produces neural excitation at a specific region of the cochlea corresponding to a characteristic frequency (tonotopy) [9]. In CIs however, a broad excitation is produced, mainly because the fluids in the cochlea are highly conductive causing the charge to spread along the cochlea. This phenomenon is commonly referred to as spread of excitation. Spread of excitation in the cochlea causes that different electrodes in the cochlea interact with each other, this is known as channel interaction. It has also been demonstrated that spread of excitation might explain some of the performance variability in speech intelligibility observed in CI users [19]. Current CI devices use monopolar stimulation mode, this means that an electrode in the scala tympani is activated and current flows to an external electrode, typically situated close to the skull or the housing of the implant. Additionally, it has been shown that the amount of spread of excitation is influenced by the anatomy and the conductivity of the tissues in the cochlea [3, 6, 8, 17]. This has been confirmed using in vivo measurements of the cochlea with an implanted electrode, using resistive models and solving analytical equations of the 3D volume conduction problem.

Resistance networks introduced by von Békésy (1951) as transmission line models, have been used to calculate the voltage in the scala tympani as a function of distance from the cochlear base. It has been reported that these type of models are of value in the estimation of current interactions but they do not provide with the resolution necessary to simulate individual excitation process. In order to obtain more accurate simulations, boundary element method (BEM) and finite element method (FEM) have been proposed. Finley et al. [6] were the first to present an integrated 3D neuron field model of a segment of an unrolled cochlea, using the FEM. In [7, 8] they presented a rotationally symmetric cochlear geometry to calculate neural excitation patterns for different electrode configurations and stimulation patterns. In [3] the model was improved using a more refined helical representation of the cochlea. In [10] an FEM model based on a spiral shaped cochlea was presented which was used to model spread of excitation with simultaneous electrode stimulation. In [16] a simplified spiraled model of the human cochlea was developed from a cross sectional microphotography. More recently in [17] a three-dimensional FEM model of the cochlea was developed to obtain the voltage distribution at positions closer to the site of neural stimulation. This model was used to demonstrate the way the voltage distribution varies with the geometry of the cochlea and the electrode array.

As mentioned above, current commercial CI systems use monopolar stimulation. This produces a broad voltage distribution in the cochlea causing large channel interaction. Different stimulation modes have been proposed to try to compensate the negative effects of such channel interaction. For example, bipolar and tripolar stimulation can be used. The idea of these methods is to narrow the electrical field and produce more place-specific neural activation in the cochlea [1]. In bipolar stimulation mode two adjacent electrodes, the active and the reference electrode are used to focus the electrical field. In tripolar stimulation mode three intracochlear electrodes are activated simultaneously. One electrode acts as the active electrode and the two adjacent electrodes are used as reference electrodes. Although these stimulation modes are very promising, so far it has not been possible to show benefits in terms of sound quality and speech intelligibility in CI users [1]. One possible reason to explain these results is the fact that the amount of spiral ganglion cell survival and the degree of dendrite degeneration might also contribute to sound perception. For example, if a narrower field is applied to a region of the cochlea with a low number of active neurons, the current will have to be increased to reach neighboring neurons and the spread of excitation will become automatically wider.

FEM simulations provide insight into the coupling between the electrical field and the neural response and hopefully can give more insight on understanding the individual characteristics of each cochlear implant for different stimulation modes.

In this paper, we also use a 3D finite element model of the cochlea to further study the effects of electrical spread of excitation for different electrode positions and cochlear sizes using different stimulation modes (monoplar, bipolar and tripolar). The chapter is organized as follows: first, the methodology followed to build a 3D model of the cochlea and the FEM to simulate different cochlea geometries, electrode configurations and electrode positions is presented; next the results on the simulations are given. Finally, a discussion and conclusions based on the results are given.

2 Methods

In this section we present the methodology to construct the electrical field model of the cochlea. First, the geometry of the cochlea is constructed; second, material properties to each domain of the geometry are given; and third, the electrical field is computed using FEM.

2.1 Geometry

The cochlear geometry was constructed from histological data [23] from a single human cochlea (Fig. 1). From this data we followed a very similar procedure as in [16] to create the geometry. The cochlear structure was traced using Autodesk Inventor ©. In a first step, the shapes of the compartments were approximated by polygons with a relatively low number of key points (Fig. 3). In a second step the same structures were repeated in planes every 30° around the vertical axis (Y) fitting the histological data. As a result, we obtained a cochlea similar to the one presented in (Fig. 3).

One of the advantages of the method used to generate the cochlea is that it can be resized and adapted to individual cochlear shapes. Figure 3b presents a resized cochlea adapted to clinical CT scans of the temporal bone before implantation. The cochlear resizing is achieved by adapting the largest distance from the round window to the lateral wall (distance A) and the perpendicular distance (B) as defined in [5].

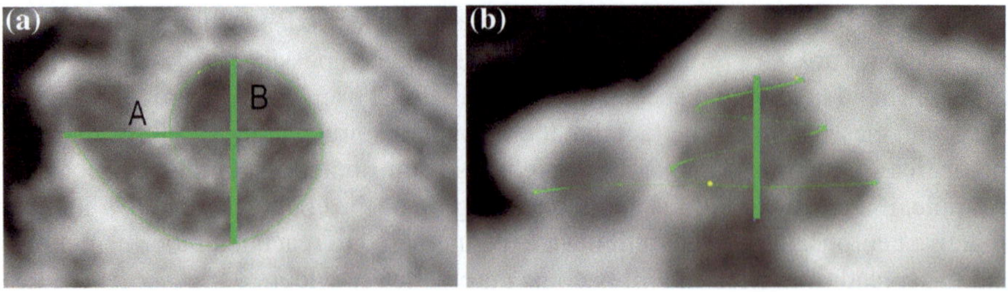

Fig. 1 **a** Cochlear dimensions estimated from clinical computer tomography. *A*- and *B*- value in CBCT data as clinical metric parameters. **b** Height of the cochlea starting from the lowest basal point to the apex

Table 1 Cochlear dimensions for the cochleas "C1" and "C2"

	0.5 TL	1 TL	1.5 TL	2 TL	2.5 TL	A-value (mm)	B-value (mm)	Height (mm)
C1	11.61	22.39	32.44	39.02	43.95	7.74	5.21	6.74
C2	14.05	25.27	35.44	43.08	48.03	8.43	5.93	5.37

Additionally the cochlear length at different places (0.5, 1.5, 2, 2.5 turns of the cochlea) and the height of the cochlea can be adapted from clinical cone beam computed tomography (CBCT) data (Fig. 1a, b). As an example, the dimensions for two cochleas ("C1" and "C2") are given in Table 1.

Next, a CI electrode array was modelled using ball/sphere electrodes positioned inside the scala tympani. This simple approach was taken because it allows for an easy parameterization of experiments by simply shifting the positions of the electrodes. Additionally, the positions of the electrodes can be adapted to the individual position of each CI user by analyzing post-operative CT scans (Fig. 2). Each electrode was placed at a distance of 1.5 mm with respect to the next electrode. In total 16 electrodes were placed inside the scala tympani.

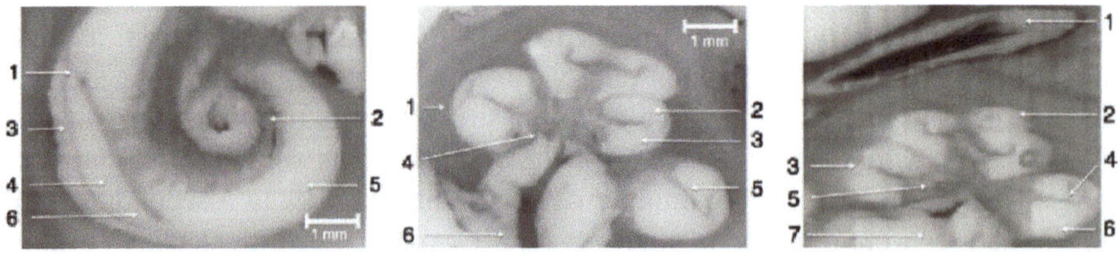

Fig. 2 *Left* Cochlea in sagital cross-section. *1* Basilar membrane, *2* Modiolus, *3* Spiral ligament, *4* Scala vestibuli, *5* Scala tympani, *6* Limbus laminae spiralis *Center* Cochlea in axial cross-section. *1* Spiral ligament, *2* Scala vestibuli, *3* Scala tympani, *4* Modiolus, *5* Osseous spiral lamina in contact to the basilar membrane, *6* Cochlear nerve in internal auditory meatus *Right* Cochlea in sagital cross-section *1* Facial nerve in facial nerve canal, *2* Scala vestibuli, *3* Spiral ligament, *4* Osseous spiral lamina in contact to the basilar membrane, *5* Modiolus, *6* Scala tympani, *7* Cochlear nerve in internal auditory meatus

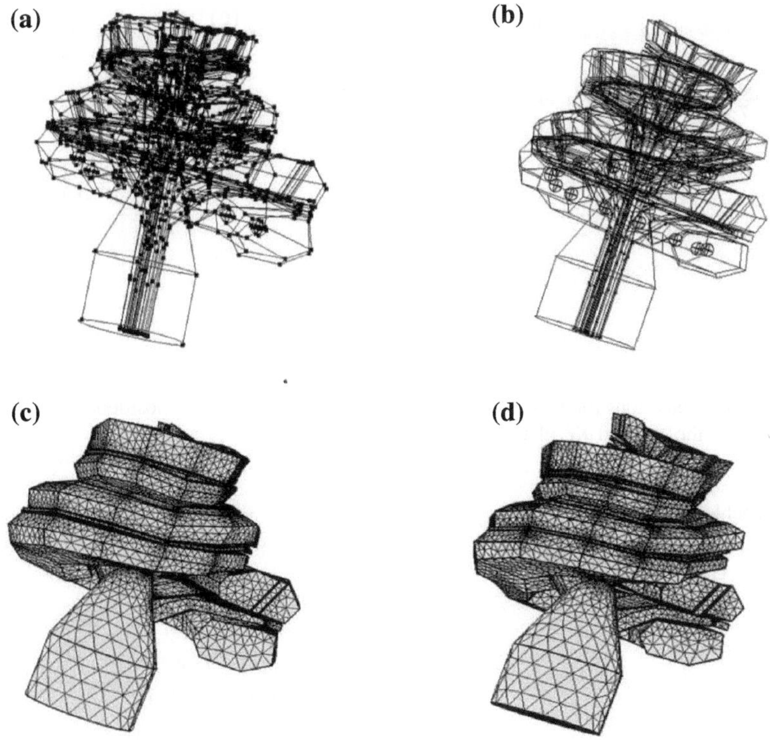

Fig. 3 Computed Aided Design (CAD) models and generated meshes adapted to the cochlear dimensions of two patients based on clinical CT images. Figures on the left correspond to the cochlea "C1" and figures on the right correspond to the cochlea "C2"

The cochlear geometry was discretized into elements with a minimum size of 0.2 mm leading to a total of 485654 and 441389 elements for the cochlea "C1" and "C2" respectively (Fig. 3c, d).

Along the CAD model of the cochlea, measurements of the electrical field were performed at the electrode positions, and at different positions of the spiral ganglions. Each spiral ganglion was modeled by 10 nodes (Fig. 4a). In total 3,000 spiral ganglions positions were modelled covering two and half turns of the cochlea as presented in Fig. 4.

2.2 Materials

The geometry was classified into domains, and each domain was assigned a material property in the form of conductivity. The conductivity values are listed in Table 2.

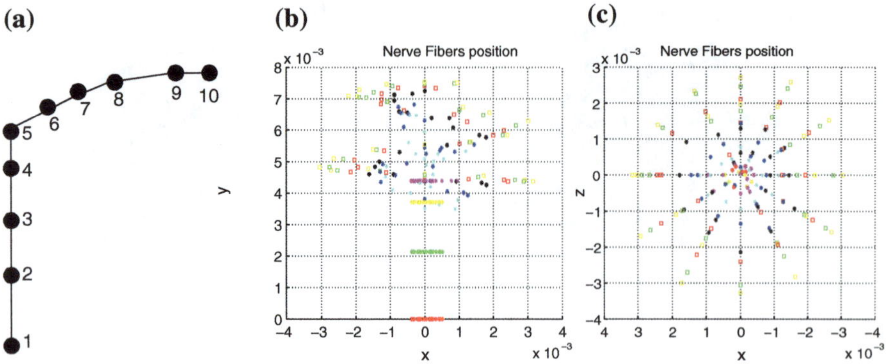

Fig. 4 **a** Node positions for a single spiral ganglion inside, **b** and **c** node positions for all the spiral ganglions for the XY and XZ planes respectively

Table 2 Conductivity values for the cochlea structures [3]

Structure	Conductivity Sm^{-1}
Electrodes	1000
Scala media	1.67
Scala tympani	1.43
Scala vestibuli	1.43
Basilar membrane	0.0625
Reissner membrane	0.00098
Nerve	0.3
Bone	0.156

2.3 *Finite Element Method (FEM)*

FEM for the given geometry was solved using the COMSOL solver ©. In this paragraph we briefly review the electrostatic equations solved by the FEM. The different elements in which the geometry was decomposed are reconnected at nodes. Next the electric field is approximated at these nodes resulting in a set of algebraic equations. The basis for the equations is the Poisson equation which gives the electrostatic potential for a given charge distribution:

$$-\varDelta u = f, \quad f \epsilon \varOmega \tag{1}$$

where $\varDelta$ is the Laplacian operator given by:

$$\varDelta = \frac{\partial^2}{\partial x_1^2} + \frac{\partial^2}{\partial x_2^2} + \cdots + \frac{\partial^2}{\partial x_d^2}. \tag{2}$$

And Ω is a d-dimensional domain (volume in 3d). The unknown function is the electrostatic potential u and the given data is the charge distribution f.

Next, the Gauss' law, also known as Gauss flux theorem, which is a law relating the distribution of electric charge to the resulting electric field is used. In its integral form, the law states that, for any volume V in space, with boundary surface ∂V, the following equation holds:

$$\int_{\partial V} \mathbf{E} \cdot ndS = \int_{\partial V} \frac{\rho}{\epsilon} \cdot dV = \frac{Q_v}{\epsilon}, \tag{3}$$

where the left hand side is called the "electric flux" through ∂V, $\mathbf{E}$ is the electric field $\int_{\partial V} ndS$ is a surface integral with an outward facing surface with a normal orientation. The surface ∂V is the surface bounding the volume V, Q_V is the total electric charge in the volume V, ϵ is the electric constant - a fundamental physical constant. Again using divergence theorem, the differential form of the Gauss'law is:

$$\nabla \cdot \mathbf{E} = \frac{\rho}{\epsilon}, \tag{4}$$

where ρ is the charge density. Since the line integral of the electric field around any close curve in space equals 0 ($\oint E \cdot dl = 0$),

$$\mathbf{E} = -\nabla u. \tag{5}$$

Eliminating by substitution, we have a form of the Poisson equation:

$$\Delta u = -\frac{\rho}{\epsilon}. \tag{6}$$

In practice $\mathbf{E}$ can be calculated as the difference between the elecric potential at two positions divided by the distance between those positions as follows:

$$\mathbf{E} = \frac{\partial u}{\partial x} \cong \frac{u_1 - u_2}{\Delta}. \tag{7}$$

Therefore, if we know the electric field, we can calculate the voltage at different points. The current density in the medium $\mathbf{J}$ can be obtained from the Ohm law as follows:

$$\mathbf{J} = \sigma \mathbf{E}, \tag{8}$$

where σ is the electrical conductivity of the material in the model (S/m).

2.4 Individual Model of a Degenerate Nerve Fiber

It has been shown that in persons with profound hearing loss, the dendrites are almost lost but the somas and axons survive, this is known as retrograde neural degeneration. It has been also shown that the degree of retrograde concerns only Type I auditory nerve fibers [20]. A degenerate version of the auditory nerve model similar to the one presented in [20] was used to simulate its effect. This was modelled by removing the first four nodes of each nerve presented in Fig. 4a.

3 Results

3.1 Experiment 1: Simulation of the Voltage Distribution for Different Electrode Positions

In the first part of the experiment we simulated the intracochlear voltage at different electrode positions. Current CI devices are equipped with measurement circuitries that provide enough resolution to capture the intracochlear potential at the electrode contacts. This objective measurement is known as electrical field imaging (EFI) [12]. In an EFI measurement current is typically flowing between one intracochlear electrode contact and an extracochlear reference located near the housing of the implanted stimulator (although the reference electrode position depends on cochlear implant manufacturers). This monopolar stimulation forces the current to spread through all cochlear tissues. The details of the intracochlear potential profiles, therefore, depend on the anatomy and conductivities of the tissues and on the properties of the electrode contacts. These factors differ from subject to subject and may vary over time [11]. The finite element model of the electrically stimulated cochlea was used to simulate an EFI measure. Monopolar stimulation was simulated stimulating one active electrode within the array in the cochlea, and defining the reference electrode as a cube of size $50 \times 50 \times 50$ mm that houses the cochlear implant. Three different electrode insertions/positions in the scala tympani were simulated (Fig. 5): (1) Modiolar electrode position where the electrode is placed close to the nerve, (2) midscalar electrode position where the electrode array is placed in the middle of the scala tympani, and (3) lateral electrode position where the electrode array is placed close to the lateral wall of the cochlea.

Each electrode was stimulated individually and the resulting voltage was measured at the stimulating and non-stimulating electrode contacts. Figure 6 presents the results of the simulations for the three electrode positions. As expected, the model simulations show two distinct areas around each electrode being stimulated [2]. The first being the reactive component originating beyond the stimulated electrode. This region shows voltage distribution that decays exponentially with electrode distance. The second region corresponds to the reactive component originating at the electrode nerve interface of the electrode being stimulated. It has been shown that the voltage

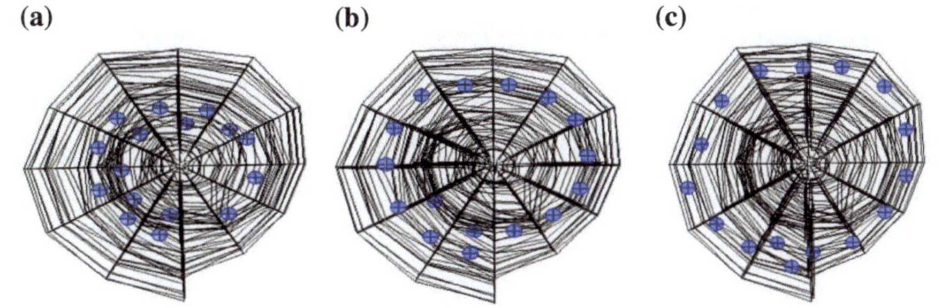

Fig. 5 Different electrode positions in the scala tympany on the XZ plane

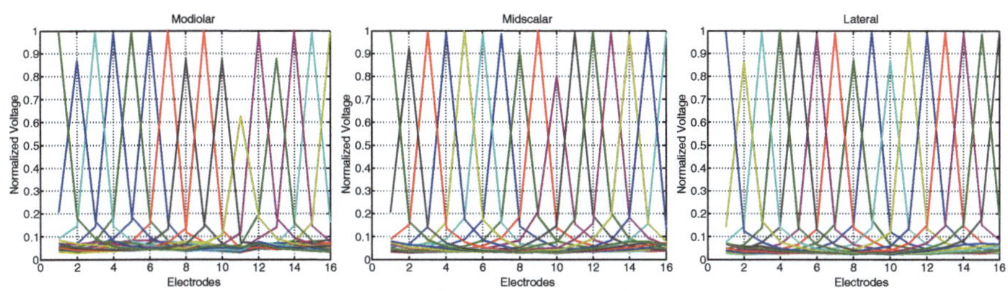

Fig. 6 Electrical field imaging simulations for three different electrode positions (Modiolar, Midscalar and Lateral). Each different colored curve corresponds to excitation of a single electrode, given by the position of the peak voltage in the electrode array

within this region decays lineraly. Because the reactive component is quite large it is difficult to observe large differences in the EFI measures for the different electrode placements without making additional model assumptions.

In the second part of the experiment each electrode was stimulated individually and the resulting voltage distribution was then measured along the assumed position of the spiral ganglion pathway for nodes 10, 8 and 7 presented in Fig. 4a. Figure 7 presents the results of the simulations and Table 3 presents the average width of the spread of excitation for each node and electrode position. The width is measured as the distance between the two points placed at 75 % with respect to the maximum of the extracellular voltage. The mean width over all electrodes for the different nodes has been calculated and can be observed in Table 3.

From Table 3 it can be viewed that the width of the spread of excitation depends on the electrode placement and on the preservation of the auditory nerve. If there is no auditory nerve degeneration (Node 10), modiolar placement produces wider spread of excitation than midscalar or lateral electrode insertion. However, the opposite trend is observed for low nerve degeneration (Node 7). This result might be surprising, but actually it is explained by the fact that for lateral electrode position, the electrodes are placed closer to node number 10 than in modiolar position producing a narrower spread of excitation.

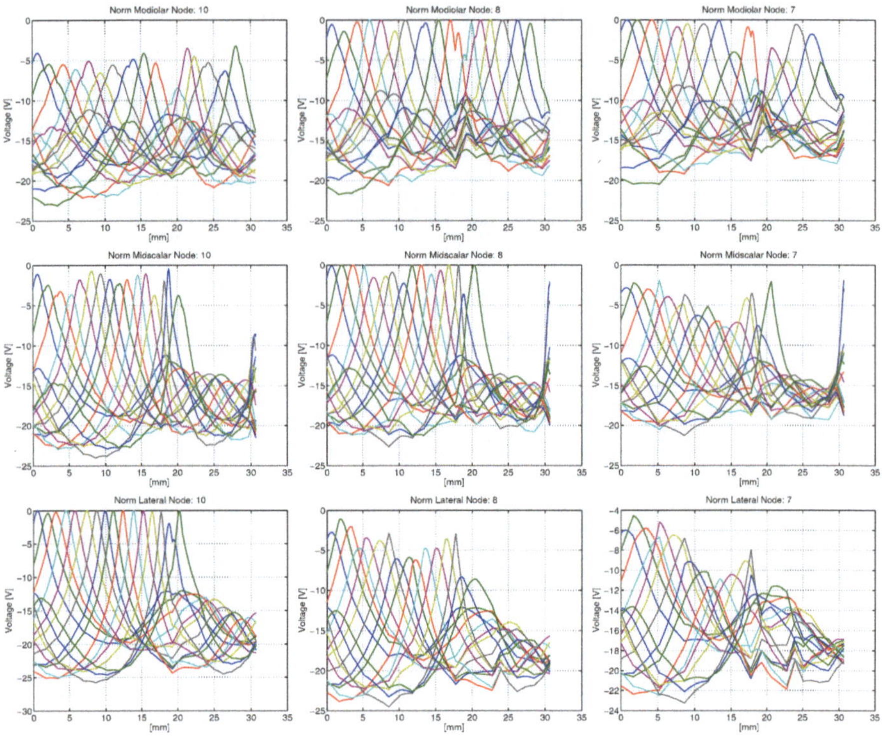

Fig. 7 Extracellular voltage at nodes 10, 8 and 7 when electrodes 1 to 16 are stimulated. For node 10 (no nerve degeneration) only small differences between electrode positions can be observed. For node 8 (*medium degeneration*) differences can be observed. For node 7 (*large degeneration*) less differences can be observed. Note that the graphs here presented have been normalized for each electrode position (modiolar, midscalar and lateral). Each color represents a different electrode being stimulated

Table 3 Spread of excitation width in millimeters at different nodes for different electrode positions

	Node 10 (mm)	Node 8 (mm)	Node 7 (mm)
Modiolar	2.25	1.55	2.2
Midscalar	1.30	2.1	3.05
Lateral	1	1.8	4.2

The model was also used to simulate the voltage decay at the spiral ganglion (for nodes 10, 5 and 1) for different placements within the scala tympani for electrode 1 (the most basal one) as presented in Fig. 8. The electrode was moved from a modiolar position to a lateral position and the voltage was measured at the different nodes. As expected, the placement of the electrode array closer to the modiolus (and therefore, closer to the nerve) fibers, produces higher voltage through the spiral ganglion cells.

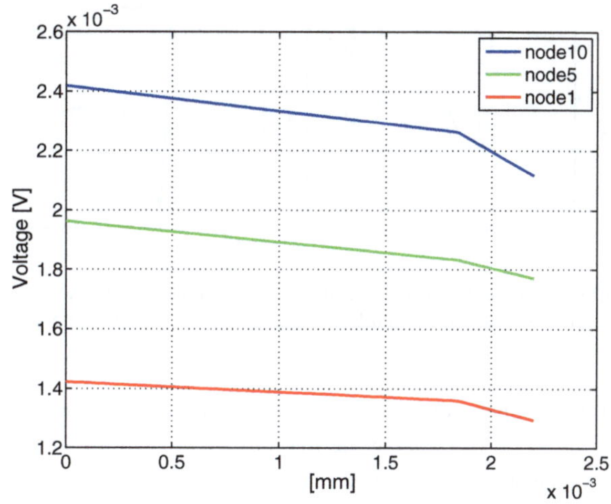

Fig. 8 Voltage (*y-axis*) decay for different electrode positions inside the scala tympany (*x-axis*). 0 mm means that the electrode is touching the internal wall of the scala tympani. The simulation was performed at nodes 10, 5 and 1

Additionally, an electrode array positioned close to the modiolus means that the most apical electrodes will be more deeply inserted covering a larger area of spiral ganglion cells. These simulations show that modiolar electrode placement permits the stimulation of more apical frequency regions (low frequencies). From Fig. 7, it can be observed that the extracellular voltage presents a secondary peak with lower amplitude produced by cross turn stimulation. This observation was already presented in [17]. In summary, these simulations show large differences in extracellular voltage measured at the spiral ganglions for different electrode positions. Additionally, these differences become larger for more degenerated auditory nerves. The large differences in extracellular voltage could give some insight in explaining the inter-subject variability in speech intelligibility observed in cochlear implants users, which cannot be explained by EFI measures.

3.2 Experiment 2: Simulation of the Voltage Distribution for Different Stimulation Modes—Monopolar, Bipolar and Tripolar

The FEM was also used to simulate different stimulation configurations: monopolar, bipolar and tripolar stimulation. For monopolar stimulation current was injected to electrode 11 and the reference electrode was outside the cochlea. For bipolar stimulation, electrode 10 and 11 were used as active and reference electrodes. For tripolar stimulation, electrode 11 was used as active electrode and electrodes 10, and 12 were used as the reference. Figure 9 presents the extracellular voltage measured at nodes 10, 8 and 7 along the spiral ganglion for monopolar, bipolar and tripolar stimulation (modiolar electrode insertion). Table 4 presents the average spread of

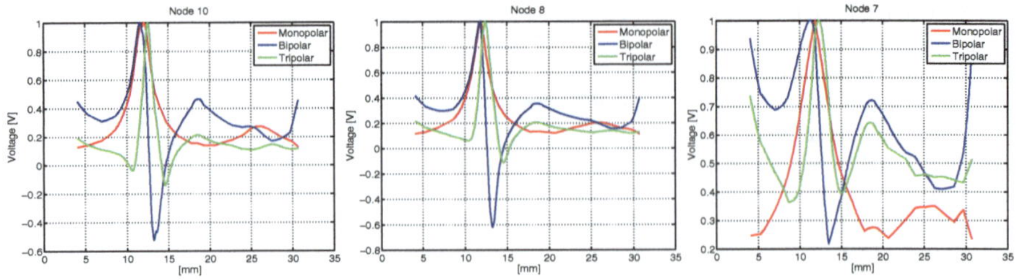

Fig. 9 Extracellular voltage at nodes 10, 8 and 7 with a modiolar electrode insertion for monopolar, bipolar and tripolar stimulation mode. For better presentation, the extracellular voltage has been normalized for each stimulation mode

Table 4 Spread of excitation width in (mm) at different nodes and for different stimulation modes

	Node 10 (mm)	Node 8 (mm)	Node 7 (mm)
Monopolar	1.78	1.3	2.79
Bipolar	1.25	0.98	3.5
Tripolar	0.78	0.9	2.1

excitation width for the different stimulation modes and for different degrees of dendrite degeneration.

From Table 4 and Fig. 9 it can be observed that for no nerve degeneration (node 10), monopolar stimulation produces wider spread of excitation than bipolar or tripolar. However, as the dendrites are more retracted (node 7) the differences in spread of excitation width for the different stimulation modes are less marked.

3.3 Experiment 3: Simulation of the Voltage Distribution for Different Cochlear Anatomies and Stimulation Modes

This experiment investigates the effect of cochlear size on extracellular voltage along the spiral ganglion. Two cochlear geometries were used: "C1" and "C2" using a mid-scalar electrode placement for both cochleae. The extracellular voltage was computed for monopolar, bipolar and tripolar stimulation modes using the same configuration as in Experiment 2. The results obtained are presented in Fig. 10. The results show differences in the shape of the excitation pattern for the two cochlear geometries. For example, for the cochlea "C1" the peaks of the extracellular voltage for monopolar and bipolar stimulation are shifted with respect to cochlea "C2". Furthermore, for the cochlea "C2" monopolar stimulation produces a narrower spread of excitation than for cochlea "C1". From these experiments, it seems that the size and shape of the cochlea can also contribute to explain the inter subject variability in speech intelligibility typically observed in CIs users.

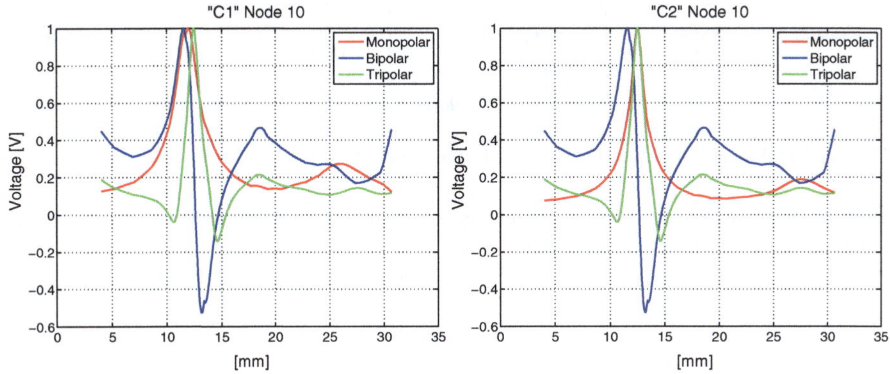

Fig. 10 Extracellular voltage for two cochlea anatomies "C1" (*left*) and "C2" (*right*) for different stimulation modes: monopolar, bipolar and tripolar

Table 5 Spread of excitation width at node 10 for two different cochlear geometries "C1" and "C2" and for different stimulation modes

	C1 (mm)	C2 (mm)
Monopolar	1.78	0.9
Bipolar	1.25	1.3
Tripolar	0.78	0.75

Again from Table 5, it can be observed that the dimensions of the cochlea have an influence on the width of the spread of excitation for different stimulation modes. For the cochlea "C2" bipolar stimulation produces a wider spread of excitation than monopolar.

3.4 Experiment 4: Automatic Instrument Identification Using the Model of the Electrically Stimulated Cochlea

We used the FEM model to simulate the excitation patterns produced by different instrumental sounds with very similar spectral characteristics. A sound coding strategy, similar to these used in commercial sound processors, is used to process the instrumental sounds and compute the currents that are applied to the electrodes in the FEM model. More details about the sound coding strategies can be obtained from [13, 14].

The extracellular voltage distribution obtained from the model is added with a white gaussian noise component. This additive noise component is necessary to increase the sensitivity of the classifier (Fig. 11). Finally, a discrete cosine transform (DCT) is applied over the noisy voltage. The reason for using the DCT is to focus the

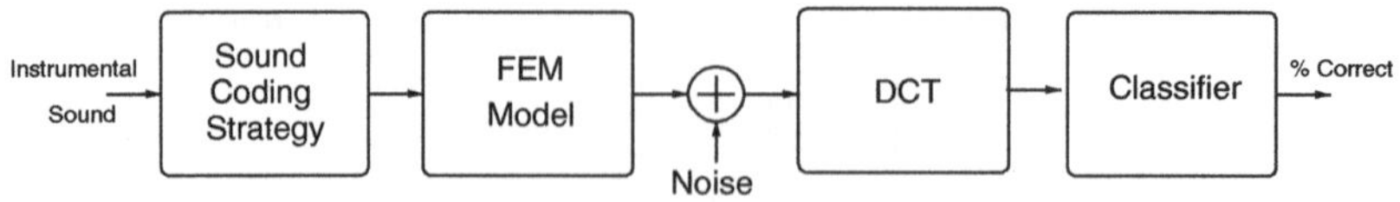

Fig. 11 Block diagram of the automatic instrument recognizer

Table 6 Results of the automatic instrument identification task for different stimulation modes and different amounts of neural degeneration

Scores (%)	Node 10 (%)	Node 8 (%)	Node 7 (%)
Monopolar	50	50	50
Bipolar	55.55	66.66	50
Tripolar	72.22	72.22	55.55

classifier on timbral features, which are more relevant for instrument identifaction. Some instrumental sounds were processed with the described method to obtain the templates that will be used as reference to perform the classification. To classify new instrumental sounds, the same process as the one described to compute the templates is followed. Next, the classifier compares the values generated by the new instrumental sounds with the stored templates. The instrument template that minimizes the dissimilarity with the new sound is selected by the class it belongs to. The similarity measure uses Dynamic Time Warping (DTW) [18]. Instrumental sounds containing a single note (B3 and C4) of a cello, viola, trumpet, trombone, flute and saxophone were used to compute the templates. The % correct classification was obtained by processing single notes of the same instruments containing the notes C3 and B4. That means that 12 instrumental sounds were used for the evaluation. Table 6 presents the results of this experiment which serves to illustrate the variability in the identification of sounds for different stimulation modes and different amounts of the neural degeneration.

4 Discussion

This manuscript has presented the development of a model of the electrically stimulated cochlea based on FEM. The model has been developed with the goal to assess individual differences on the peripheral processing of cochlear implant users. For example the spread of excitation which is believed to contribute to speech intelligibility and sound performance in CI users has been examined in four experiments.

Experiment 1 analyzed the variability in the extracellular voltage and the width of the spread of excitation for different electrode positions (modiolar, midscalar and lateral). The model was able to predict that in general modiolar electrode position is able to produce narrower voltage distribution if there is auditory nerve degeneration.

However, in case of complete preservation of the audiotory nerve, modiolar electrode positions could not reduce the spread of excitation.

In experiment 2, different stimulation modes were simulated. The model predicted that monopolar stimulation produces a wider spread of excitation than bipolar and that bipolar stimulation produces a wider spread of excitation than tripolar for non-degenerated auditory nerves. However, for degenerated auditory nerves the differences between the three stimulation modes became much smaller. This means that depending on the electrode-nerve interface of each individual some stimulation modes can be more beneficial than others.

Experiment 3 investigated the differences in extracellular voltage for different cochlear anatomies. The model presented here is based on existing models and although simple, it allows for a relatively easy adaptation to clinical CT images of CI individuals. An example of two different cochleas has been presented. From the simulations it was observed that the width of the spread of excitation depends on the anatomy of the cochlea.

The scope of the manuscript is focused on the development of the model. The model should serve as a tool in the future such as to understand better the variability of each individual. The validation of the model, comparing the simulations with real data is out of the scope of this paper.

Finally, experiment 4 was designed to illustrate the impact of peripheral aspects (such as the electrical field or the amount of neural degeneration) on the perception of sounds. The goal was to demonstrate that at least for computers, differences in the excitation patterns provided to the cochlear implant can have an influence in sound identification. For example, we could show that stimulation modes causing a narrower spread of excitation (tripolar stimulation) obtained higher identification scores than stimulation modes that produce a wider spread of excitation (monopolar stimulation).

5 Conclusions

A model of the electrically stimulated cochlea has been developed. This model has been used to assess the impact of electrode position, cochlear anatomy and different stimulation modes on the spread of excitation. Simulations show variability in the spread of excitation for all these configurations of the model. An automatic sound classifier was able to illustrate the effects in correct classification rate for the different configurations of the CI electrical model.

Acknowledgments The authors would like to thank the valuable contributions to the development of the model from Prof. Antoni Ivorra, Marcel Farres and Nikos Papachristou from the Universitat Pompeu Fabra. This work was supported by the DFG Cluster of Excellence EXC 1077/1 "Hearing4all".

References

1. Berenstein, C.K., Mens, L.H., Mulder, J.J., Vanpoucke, F.J.: Current steering and current focusing in cochlear implants: Comparison of monopolar, tripolar, and virtual channel electrode configurations. Ear Hear. **29**, 250–260 (2008)
2. Berenstein, C.K., Vanpoucke, F.J., Mulder, J.J., Mens, L.H.: Electric field imaging as a means to predict the loudness of monopolar and tripolar stimuli in cochlear implant recipients. Hear. Res. **270**, 28–38 (2010)
3. Briaire, J.J.: Cochlear implants from model to patients, Thesis, ISBN 978-90-9023555-4, Universiteit Leiden (2008)
4. Colletti, L., Mandal, M., Colletti, V.: Cochlear implants in children younger than 6 months. Otolaryngol. Head Neck Surg. Off J. Am. Acad. Otolaryngol. Head Neck Surg. **147**(1), 139–146 (2012)
5. Escudé, B., James, C., Deguine, O., Cochard, N., Eter, E., Fraysse, B.: The size of the cochlea and predictions of insertion depth angles for cochlear implant electrodes. Audiol. Neurotology **11**(suppl 1), 27–33 (2006)
6. Finley, C., Wilson, B., White, M.: Models of neural responsiveness to electrical stimulation. In: Miller, J., Spelman, F. (eds.) Cochlear Implants, pp. 55–96. Springer New York (1990)
7. Frijns, J.H.M., de Snoo, S.L., ten Kate, J.H.: Spatial selectivity in a rotationally symmetric model of the electrically stimulated cochlea. Hear. Res. **95**, 33–48 (1996)
8. Frijns, J.H.M., de Snoo, S.L., Schoonhoven, R.: Potential distributions and neural excitation pattervs in a rotationally symmetric model of the electrically stimulated cochlea. Hear. Res. **87**, 170–186 (1995)
9. Greenwood, D.D.: A cochlear frequency-position function for several species- 29 years later. J. Acoust. Soc. Am. **87**, 2592–2605 (1990)
10. Hanekom, T.: Thesis—cochlea modelling. In: Faculty of Engineering, built Environment and Information Technology. University of Pretoria, Pretoria (2001)
11. Hughes, M.L., Vander Werff, K.R., Brown, C.J., Abbas, P.J., Kelsay, D.M., Teagle, H.F., Lowder, M.W.: A longitudinal study of electrode impedance, the electrically evoked compound action potential, and behavioral measures in nucleus 24 cochlear implant users. Ear Hear. **22**, 471–486 (2001)
12. Mens, L.H., Boyle, P.J., Mulder, J.J.: The Clarion electrode positioner: Approximation to the medial wall and current focussing. Audiol. Neurotol. **8**, 166–175 (2003)
13. Nogueira, W., Bchner, A., Lenarz, Th, Edler, B.: A psychoacoustic, "NofM"-type speech coding strategy for cochlear implants. EURASIP J. Adv. Sig. Process. **2005**, 101–672 (2005)
14. Nogueira, W., Litvak, L., Edler, B., Ostermann, J., Bchner, A.: Signal processing strategies for cochlear implants using current steering. EURASIP J. Advan. Sig. Process. **2009**, 213–531 (2009)
15. De Raeve, L.A.: Longitudinal study on auditory perception and speech intelligibility in deaf children implanted younger than 18 months in comparison to those implanted at later ages. Otol Neurotol **31**(8), 1261–1267 (2010)
16. Rattay, F., Leao, R.N., Felix, H.: A model of the electrically excited human cochlear neuron. II. Influence of the three-dimensional cochlear structure on neural excitability. Hear. Res. **153**, 64–79 (2001)
17. Saba, R.: "Cohlear implant modelling: Stimulation and power consumption", Thesis, university of Southampton. Faculty of Engineering and Environment, Institute of Sound and Vibration (2012)
18. Sakoe, H., Chiba, S.: Dynamic programming algorithm optimization for spoken work recognition. IEEE Trans. Acoust. Speech Signal Process. **26**(1), 43–49 (1993)
19. Shannon, R.V., Fu, Q.J., Galvin III, J.: The number of spectral channels required for speech recognition depends on the difficulty of the listening situation Acta Otolaryngol. Suppl. **552**, 50–54 (2004)

20. Smit, J.E., Hanekom, T., Hanekom, J.J.: Predicting action potential characteristics of human auditory nerve fibers through modifications of the Hudgkin-Huxley equations. S. Afr. J. Sci. **104**, 284–292 (2008)
21. Wilson, B.S., Finley, C.C., Lawson, D.T., Wolford, R.D., Eddington, D.K., Rabinowitz, W.M.: Better speech recognition with cochlear I plants. Nature **352**, 236–238 (1991)
22. Wilson, B.S., Dorman, M.F.: "Cochlear implants: A remarkable past and a brilliant future". In: Proceedings of the 9th International Conference on Cochlear Implants and Related Sciences, pp. 3–21. Elsevier Science Bv, Vienna, AUSTRIA (1996)
23. Würfel, W., Lanfermann, H., Lenarz, T., Majdani, O.: "Cochlear length determination using Cone Beam Computed Tomography in a clincal setting", Hear. Res. **316**, 65–72 (2014)

Implant Related Infections

Wolf-Rainer Abraham

Abstract The formation of biofilms on implants by bacteria is difficult to treat, life-threatening, and costly. Hence, alternatives for the prevention of biofilm infections are urgently needed. The assessment of rhythm management devices revealed colonization of 47 % by asymptomatic biofilm communities. Comparison with infected implants showed a much higher biodiversity of the infectious biofilm communities which were dominated by pathogenic *Staphylococcus* species. The results suggest that it is not essential to suppress any biofilm formation but only pathogenic bacteria species. The situation differed considerably for biofilms on dental implants. Here parts of the implant are always in a non-sterile environment and on all implants biofilm communities could be found. More than 60 different species could be identified from infected dental implants but contrary to the pacemakers no clear pathogen was found. The results indicate different mechanisms of infections requiring individual concepts for biofilm prevention on implants.

Keywords Implant · Biofilm · Microbial community · Pathogen · Asymptomatic biofilm

1 Introduction

One of the main threats for implants is the colonization of their surfaces by bacteria where they attach and form biofilms [1]. These biofilms offer protection against, e.g. noxious agents like oxidants [2, 3], antibiotics [4] or antibodies. Especially because of their superior protection biofilms can be detrimental, e.g. in wounds, on implants, teeth or catheters, hence, controlling biofilms is one of the fundamental problems in medicine. They are difficult to treat with antibiotics because of a much higher resistance of the pathogens inhabiting the biofilms [5]. Often the replacement of the implant is the only solution against these biofilm infections.

W.-R. Abraham (✉)
Chemical Microbiology, Helmholtz Center for Infection Research, Inhoffenstrasse 7, 38124 Braunschweig, Germany
e-mail: wolf-rainer.abraham@helmholtz-hzi.de

T. Lenarz and P. Wriggers (eds.), *Biomedical Technology*, Lecture Notes in Applied and Computational Mechanics 74, DOI 10.1007/978-3-319-10981-7_11

It has been estimated that 11.3 % of all patients acquire infections during medical care [6]. These so called nosocomial infections are mainly caused by pathogens organized in biofilms. About half of the cases are associated with indwelling devices. Although less common than catheter related infections, implant related infections are more difficult and more expensive to treat. The rates of implant associated infections, however, are not uniform but dependent on the type of implant, age of patient, severity of underlying disease, etc. Infection rates of 2 % for joint prostheses, 5 % for fracture-fixation devices, 4 % for pacemaker—defibrillators and 40 % for ventricular assist devices have been reported [7]. Biofilm infections cause prolonged postoperative length of stay and high health costs worldwide, e.g. catheter-related bloodstream infections cost hospitals between $5,000 and $34,000 per infection. Furthermore, 12–25 % of bloodstream infections are attributable to patient mortality [8]. For prosthetic valve endocarditis a mortality rate of 46 % for patients treated only with antibiotics and 24 % for those treated by a combination of surgery and antibiotic therapy has been reported [9].

Although clinical microbiology focuses on pathogens bacteria seldom live as pure strains alone but use to be organized in microbial communities. One of the fundamental characteristics of these communities is their work sharing and tolerance towards environmental stresses. The species diversity is huge and reflected in the large evolutionary distances between the microorganisms [10]. The fact that many microorganisms present in the human body cannot currently be cultivated in the laboratory makes things even more complicated. This limitation has nowadays been partially circumvented by the application of molecular methods which are not based on cultivation [11]. Nevertheless, the limited cultivability of bacteria prevents the characterization of the taxa by common physiological tests leaving us with almost no clue for the requirements of several species for which no isolates are available.

Several approaches have been made to prevent biofilm formation on implants or dissolve already formed biofilms and kill the bacteria. All solutions found so far are compromises and no optimal biofilm-preventing implant material or surface has yet been found. As a consequence, novel approaches like e.g. blocking the communication between bacteria or the expression of pathogenicity factors have been introduced. Understanding the consortia in biofilms and their interactions are of fundamental importance for the control and manipulation of biofilms. To achieve this goal we have to understand the interactions of microbial communities organized in biofilms within the biofilms, the surface they are attached to and the human host.

2 Species Diversity of Biofilm Communities on Implants

Colonization of surgical implants does not necessarily mean infection. Only the clinical manifestation is a clear sign that an implant became infected. Asymptomatic implants, i.e. implants not showing clinical signs of infections, are the rule. But how many of them carry biofilms which do not cause infections? What are the compositions of these biofilm, why are they not causing symptoms of infection and are they communities consisting of several species or are they monospecific biofilms? We became interested in these questions and we assessed implants explanted for malfunction but not infection. If biofilms could be detected on asymptomatic implants a comparison with similar implants explanted because they became infected would teach a lot about infectious biofilm communities. Rhythm management devices such as pacemakers were chosen for the assessment of bacterial colonization. To make sure that any bacteria present on the implants have not been introduced during implantation only devices which were implanted for at least 5 years were analyzed.

2.1 Biofilm Communities of Asymptomatic Implants

For this purpose, devices changed because of battery depletion were ideal and the explanted devices were routinely assessed for bacterial DNA. A total of 108 devices from asymptomatic patients were examined for evidence of bacterial DNA on the device and in the surrounding tissue using single strand conformation polymorphism analysis (SSCP) [12]. In 47.2 % of the patients, bacterial DNA was demonstrated on the device, which had been in place for 63.7 months in average. The SSCP analysis revealed that also all devices carrying biofilms had multi-species biofilm communities. This is in accordance from many other environmental studies that bacterial species seldom live alone but prefer to grow in communities and consortia.

Excision of the main bands of the SSCP-gels and sequence analyses of the extracted DNA allowed the identification of the main members of the biofilm communities. From 27.8 % at least one bacterial species could be identified by sequence analysis but some sequences could not be unambiguously identified due to shortness of the amplicons. Most devices showed several bacterial species, in which 16 % belongs to the genus *Pseudomonas*, 11 % to *Staphylococcus*, 10 % to *Stenotrophomonas*, 9 % to *Rhizobium*, and 7 % to *Propionibacterium* [13]. However, some rare species, e.g. *Leptotrichia* sp., *Sphingobium* sp., *Rhizobium* sp., *Halomonas* sp., or *Burkholderia ferrariae* could also be identified. Most of these species have rarely been reported from humans but are known from human microbiome analyses. One biofilm community harbored a *Lactobacillus* species and one *Klebsiella oxytoca*. All patients showing *Staphylococcus* species on their devices developed later on infection of the devices. Therefore, these samples were excluded from the analysis.

Concerning the species composition of the biofilm communities, a huge diversity was observed between different patients and no identical biofilm communities could be detected between two different patients. This highlights again the immense and complex influence the host has on the species composition of its colonizing bacteria [14]. From the available patient data no correlations between species composition and the patient could be detected. With the exception of the few cases of *Staphylococcus* species no bacterial species could be found which are known as human pathogens. However, a number of the detected bacteria are facultative pathogens. The origin of the bacteria remained unclear. It has been speculated that some of them are derived from the skin flora and the detection of *Propionibacterium acnes* seemed to support this view. However, the frequency of *P. acnes* in the biofilm communities was too low to identify the skin as main origin of the bacteria.

The lack of pathogens in these biofilms explains the asymptomatic nature of these biofilm communities. One obvious question is: what is the role of these non-virulent biofilm communities in implant infections? Are these biofilms the entry port for pathogens, do they not contribute at all to implant infections or do they even prevent the colonization of the implants by pathogens? These questions concern the functional roles individual members of these biofilm communities have in their interactions with other bacteria and the host. Answers to these questions are not easy to find but a comparison between biofilm communities from asymptomatic implants with those showing clinical signs of infections may provide some clues.

2.2 Biofilm Communities of Infected Implants

From 71 infected rhythm management devices DNA could be extracted from 95.8 % and the biofilm communities analyzed. As for the asymptomatic devices only biofilm communities were found and no mono-specific biofilms. Compared to the asymptomatic devices generally a higher biodiversity of the biofilm communities could be observed which were, contrary to the findings with the asymptomatic devices, dominated by *Staphylococcus* species. All devices with clinical signs of infection showed without exception the presence of *Staphylococcus* species. Due to the shortness of the 16rRNA gene sequences the underlying *Staphylococcus* species could not been identified. The sequence analyses identified *Staphylococcus aureus* or the closely related *S. capitis* or *S. epidermidis* as the main bacteria in all pathogenic biofilm communities [15]. This finding agreed perfectly with the microbiological analysis based on cultivation. *Staphylococcus aureus, S. epidermidis* and *S. capitis* are long known as the main pathogens in cardiovascular implant infections [16] but are these single- or a multi-pathogen infections and what is the role of the accompanying bacterial species?

To shed some light on this problem a phylogenetic analysis of the detected bacterial species were undertaken. Again a huge diversity of very different bacteria could be identified and is shown in a phylogenetic tree together with closest known species in Fig. 1. Remarkable is the large diversity of *Pseudomonas* species detected in the pathogenic biofilm communities but none of the identified *Pseudomonas* species has yet been connected with infections in humans. Again several rare bacterial species could be identified, among them *Labrenzia*, *Leptotrichia* [17], or *Sphingobium* species [18]. At this point it became interesting to compare the asymptomatic biofilm communities with the infectious ones. It turned out that in both groups *Pseudomonas* species were detected but any given *Pseudomonas* species occurred either in the asymptomatic or the infectious biofilm communities but never in both. A comparison of the occurrence of bacterial species in the biofilm communities led to the identification of species preferable present in asymptomatic biofilms

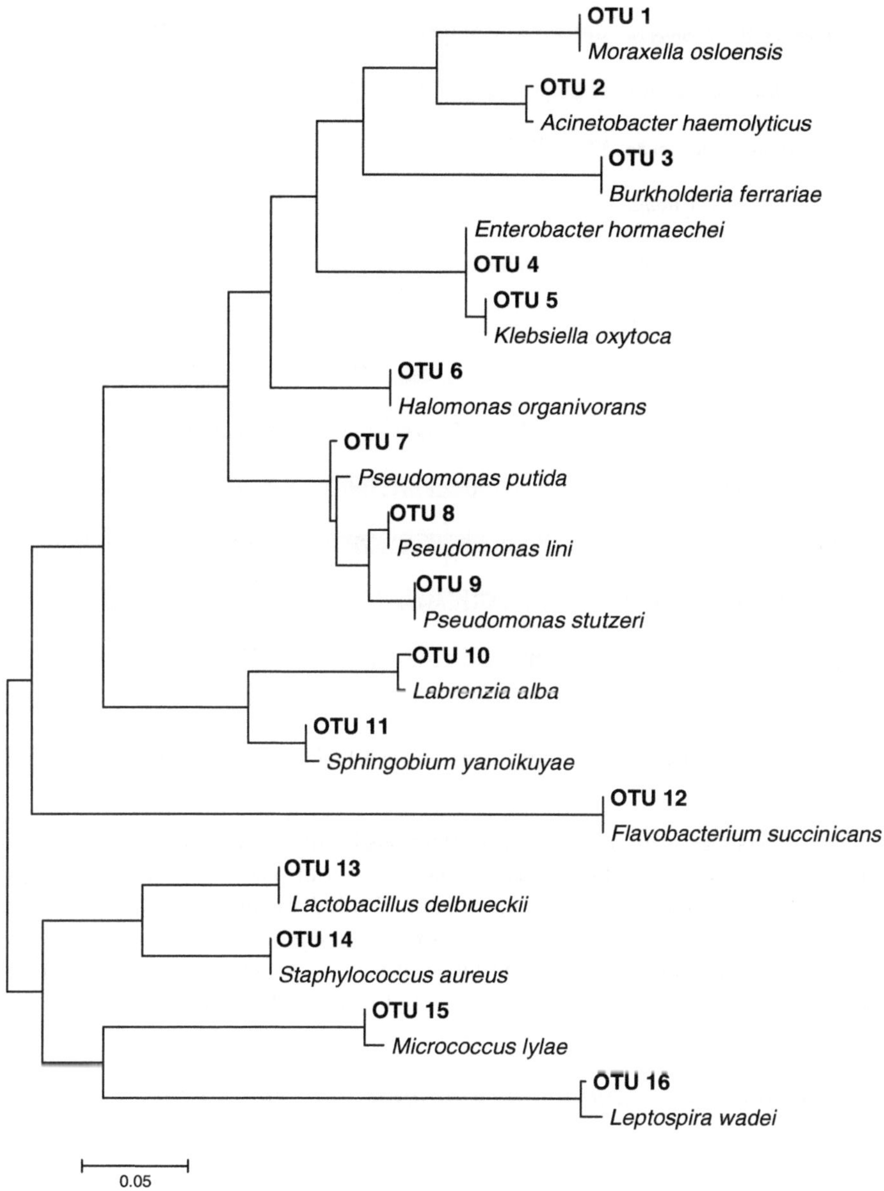

Fig. 1 Phylogenetic tree of eubacteria (OTU = operational taxonomic unit) detected in infected cardiovascular devices. The phylogeny is based on the sequences of 16 S rRNA genes and calculated using the Maximum Likelihood algorithm

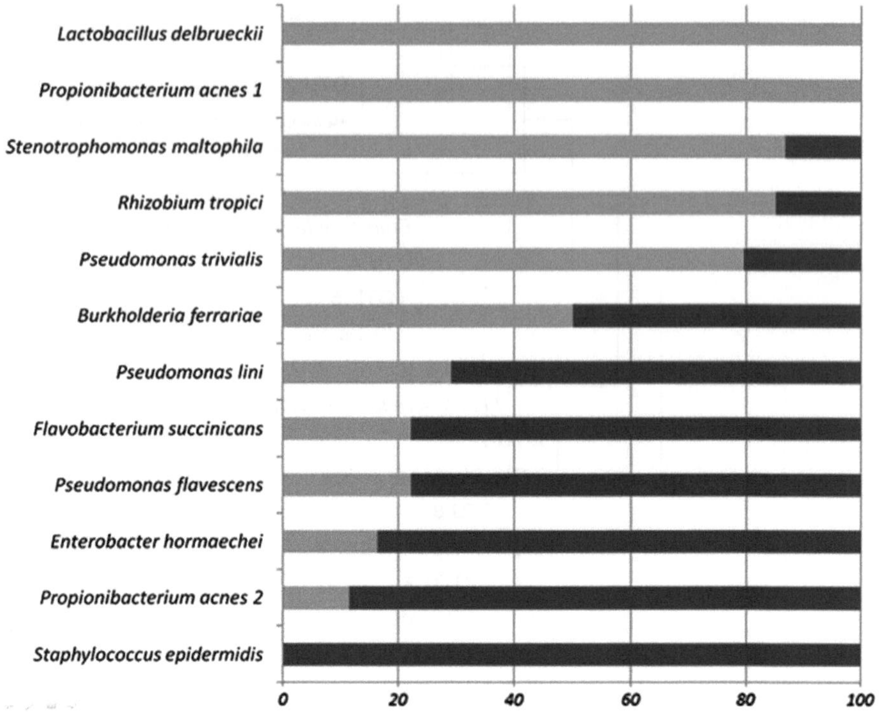

Fig. 2 The distribution of bacteria between asymptomatic (*light grey*) and infected devices (*black bars*)

while other species were indicators for infectious biofilm communities (Fig. 2). It is here interesting to note that two closely related bacterial species have been detected, here denoted *Propionibacterium acnes* 1 and 2. While *P. acnes* 1 was mainly found in asymptomatic biofilm communities, *P. acnes* 2 was a member of infectious biofilms. This observation underlines the importance of virulence factors which may play in some cases a more important role than the phylogeny of the bacteria.

2.3 Biofilms on Asymptomatic Implants in a Non-sterile Environment

The current understanding is that pacemakers or bone implants are placed into a sterile environment. But what happens to implants which are surrounded by bacteria? The skin, the gut or the mouth are always populated by a large number of bacteria, nevertheless, the mucosa and/or epithelial cells separate these microorganisms from the sterile tissues [19]. Their function as a barrier is of utmost importance for the entire organism. Implants for teeth breach this barrier and we became interested in their biofilms. As for the rhythm management devices we compared biofilm communities of asymptomatic and infected implants.

From ten patients samples of adherent supra- and subgingival periimplant biofilms were collected around titanium implants. Similar samples were also taken at remaining teeth for direct comparison of biofilms on implants and on teeth from the same patient. As for the pacemakers the microbial communities were analyzed using SSCP of 16 S rRNA gene amplicons. The diversity of *Veillonella*, *Streptococcus*, *Parvimonas*, *Fusobacterium* and *Neisseria* species detected was remarkable. The presence of various species of *Streptococcus* in the human mouth is well known and *Streptococcus mitis* and *S. intermedius* could be identified. However, no sequence belonging to the pathogen *Streptococcus mutans* was found confirming the asymptomatic nature of the biofilms. The genus *Fusobacterium* comprises strictly anaerobic bacteria and species of this genus are well known from the human mouth where they are important members of biofilm communities [20]. The species *Fusobacterium naviforme* and *F. canifelinum* were identified from the corresponding SSCP amplicons. Further well known members of asymptomatic dental biofilms, *Neisseria mucosa* and *Parvimonas micra,* were also detected in this study. As expected, a high diversity of bacteria was observed between biofilms of different patients [21]. Similar to the situation in the gut each individual seems to have in its dental biofilms its own microbial community. Applying multivariate analysis, however, revealed some bacteria species characteristic for subgingival or for supragingival biofilms. No discrimination, however, could be achieved between microbial species colonizing biofilms on implants or on residual teeth. This finding supports the view that biofilm communities on implants are not much different to the natural ones on teeth [22].

2.4 Biofilms on Infected Dental Implants

Colonization of the implants by biofilms lead to peri-implantitis, an inflammation of peri-implant tissues, and is currently seen as one of the main causes for early implant failure. It is clinically characterized by inflammation of the mucosa and a subsequent destruction of the bone around the implant. To determine the differences in community composition between healthy and infected implants a similar analysis was performed for biofilm communities from patients with infected implants. But not only samples from the site of peri-implantitis but, where possible, also samples from asymptomatic implants and residual teeth from the same patients were taken and the biofilm communities analyzed.

As for the rhythm management devices a much higher diversity in bacterial species were detected in the pathogenic biofilm communities compared to their asymptomatic cousins (Fig. 3). Remarkable is here that species known for their anaerobic lifestyles dominated the biofilm communities indicating significant changes in the physiology of the biofilms. Socransky et al. assessed a large number of microbial communities from plaque samples and correlated them with clinical data. Using multivariate analyses they grouped the bacteria into five groups increasingly related to periodontitis [23]. These groups have also been applied in the assessment of peri-implantitis although its pathogenesis is somewhat different to the one of periodontitis. In our

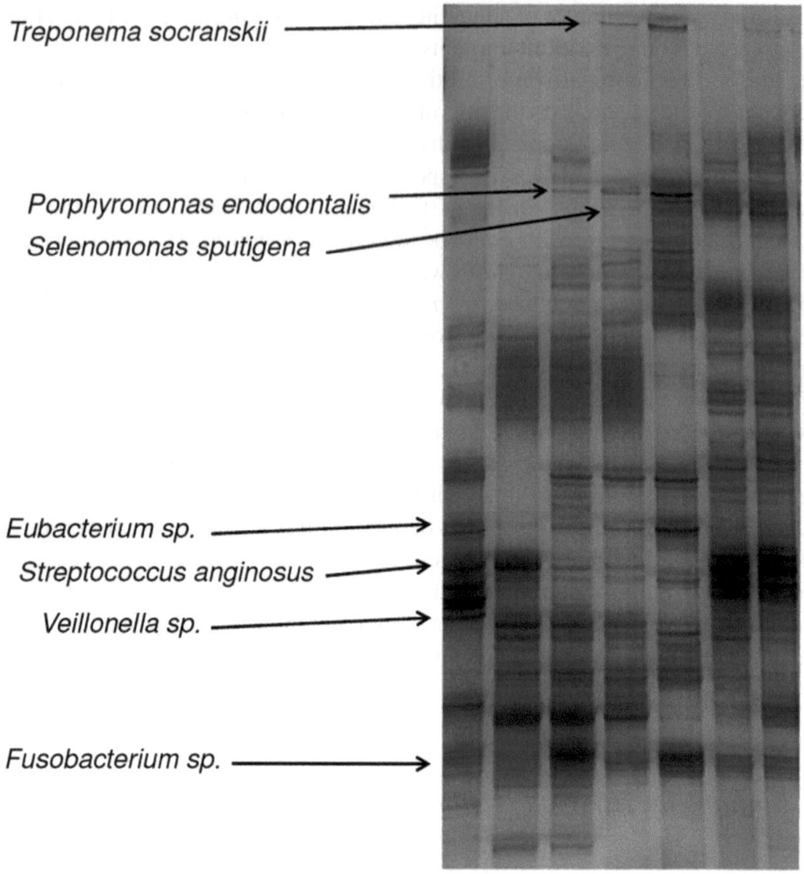

Fig. 3 Single strand conformation polymorphism (SSCP) gel of 16 S rRNA gene amplicons of biofilm communities from patients with peri-implantitis. Each lane stands for one biofilm community, each band is one distinct bacterial species. Bands migrating to the some height are identical species. Main bands have been cut out, their DNA sequenced and the species identified. Some of the identified bands are shown

study no member of the bacterial group most closely related to periodontitis was observed but in some biofilm communities *Streptococcus mutans*, known to be the leading pathogen in the formation of dental cavities, was detected. In general was the diversity of *Streptococcus* and *Fusobacterium* species in peri-implantitis biofilms much higher than in asymptomatic ones. Comparing biofilm communities between different patients led again to the impression that each patient has his own set of dental biofilm communities. Surprising, however, was the observation that no large differences in the biofilm communities between infected and healthy implant of the same patient was found. The interpretation that pathogenic biofilm communities spread over all the teeth of an individual was supported by the finding that microbial communities from infected implants and healthy residual teeth were also similar. This leaves the question open why there is only infection on one implant and none on the others although the microbial communities are more or less the same. One interpretation could be that the barrier function of the gingival has been damaged and bacteria are reaching now areas where they cause inflammation and severe immune responses. A similar scenario has been proposed as the cause for inflammatory bowel disease [24].

3 Are Asymptomatic Biofilm Communities the Beginning of an Infection?

Our findings indicate that the notion that implants are essential sterile seems not be always the case. However, the presence of biofilm communities on implants does not cause *per se* a problem because most implants did not show any clinical signs of infection and the microbial communities did not harbour any human pathogens. But, what is the role of these asymptomatic biofilm communities and—more important—are they the entry port for any implant infection? Such a question is difficult to answer, but some observations may indicate that asymptomatic biofilm communities are not the beginning of an infection. No pathogenic bacteria were found among them excluding the possibility that they are hideaways for pathogens waiting for their time to come. Furthermore, the comparison between asymptomatic and infectious biofilm communities shows fundamental changes in community composition between these types which seems to indicate that pathogens have their own group of accompanying bacteria different from those found in asymptomatic biofilms. On rhythmic management devices we found few examples of biofilms which contained pathogens but had not yet developed an infection (they all did, however, after the replacement of the implants). These biofilm communities may be seen as somehow transient to infectious one and they harboured non-pathogens which were similar to the asymptomatic biofilm communities. This may point to an invasion of asymptomatic biofilm communities by pathogens but there were too few samples to draw any clear conclusion.

The focus in clinical microbiology is on the identification of the pathogen. This makes absolutely sense because the agent causative for the disease has to be identified as fast as possible and a remedy for its control has to be found to cure the patient. But this approach ignores the non-pathogens and hampers our understanding of their role in the infection process. Only in recent years it became obvious that bacteria are present in many infectious biofilm communities which are not known as human pathogens and which do not cause infections in humans. However, it is reasonable to assume that they are no bystanders but influence, in one way or another, the progress of the infection and its course. An impressive example how pathogens can interact in biofilms is the occurrence of *Burkholderia cepacia* and *Pseudomonas aeruginosa* in human cystic fibrosis lungs. *B. cepacia* always occurs together with

P. aeruginosa in these lungs while *P. aeruginosa* is often found without a *B. cepacia* co-infection. The reason for this dependence is that *B. cepacia* reacts to the specific homoserine lactones produced by *P. aeruginosa*, but *P. aeruginosa* does not respond to the homoserine lactones produced by *B. cepacia* [25]. In this way these two pathogens manage to communicate and as a consequence to increase their resistances against antibiotics. For the role of species in asymptomatic biofilm communities one may even speculate that some of these bacteria may be useful for controlling aggressive pathogens or reduce their pathogenicity. Currently, such ideas are mere speculations because we have no clue on the functions of these non-pathogens but our understanding of mechanisms in microbial ecology has advanced considerably and the methods for the assessment of microbial interactions have been developed which will allow us the elucidation of the roles of these non-pathogenic bacteria.

Unfortunately, the situation is even more complicated because not only the interactions between the different bacterial species have to be deciphered but also their interaction with the host. That there are a multitude of interactions between the host and his bacteria has been shown in a number of impressive studies. This has also been underlined simply by the fact that individuals, even monozygotic twins, have their own characteristic microbial consortia [26]. It becomes more and more apparent that there are a huge number of interactions between our body and our microbiome and it can be predicted that unravelling the various interactions in pathogenic biofilm communities will be a slow process. But from the results already reported some conclusions for improved control of biofilm infections can be drawn.

4 Conclusion

Biofilm infections are difficult to treat due to their antibiotic resistances and a deeper understanding of the risk factors for their formation is urgently needed. To achieve this, bacterial biofilm communities of rhythm management devices from asymptomatic patients were assessed and compared with those found on symptomatic ones. From 108 devices in asymptomatic patients 47 % had bacterial DNA and remained asymptomatic. The identified bacterial taxa were untypical for clinical device infections. No correlations were found between known risk factors for device infections and the bacteria detected. The results were compared with rhythm management devices explanted due to acute infections revealing very different biofilm communities between these two types. Symptomatic biofilms were dominated by *Staphylococcus* spp. and had a much higher bacterial diversity. The results suggest that it is not essential to suppress any biofilm formation but only pathogenic bacteria species characteristic for symptomatic biofilms. Such an approach would be different to the current one where any bacterium attaching to an implant is fought independently of its pathogenicity. While all attempts to find a surface bactericidal to all pathogens turned up empty handed attempts for control of specific pathogens may be more successful.

The results reported above were then compared with asymptomatic and symptomatic biofilms from dental implants. Peri-implantitis is a serious infection of dental implants with incidence rates of up to 30 % and may cause peri-implant bone loss and implant failure. As for the rhythm management devices the biofilms from asymptomatic implants had a broad diversity between patients. No obvious pathogen could be found. From 10 patients with peri-implantitis, the sulcus fluid of the infected implants and if available those of healthy implants and residual teeth were sampled. From the biofilm communities more than 60 different species could be identified revealing a large diversity of bacteria involved in peri-implantitis. Contrary to the rhythm management devices no clear pathogen was found common to most of the infections. The finding of very diverse microbial communities in peri-implantitis points to a breakdown of the barrier function of the gingiva causing the disease. Tipping the balance in favor to epithelial cells in their "race to the surface" against bacterial cells may be here one promising approach.

The conclusion of these studies is that there is not one type of biofilm but many as there is not one biofilm pathogen but many. Tailoring implants to the specific biofilm communities expected at the implantation sites seems to be the best way to control implant related biofilm infections. Furthermore, the results point to a strong influence of the host on biofilm communities and it can be speculated that implant material specific to the needs of the individual patient may also be an option in the future.

Acknowledgments The excellent technical assistance of several members of the research group Chemical Microbiology is gratefully acknowledged. I thank Maximilian Pichlmaier (formerly Medical School Hannover, now University Hospital Munich, Germany) and Meike Stiesch (Medical School Hannover) for contributing the clinical samples and patient data to this study. Parts of the study have been supported by the German Research Foundation (DFG-SFB Collaborative Research Center 599).

References

1. Wolcott, R., Costerton, J.W., Raoult, D., Cutler, S.J.: The polymicrobial nature of biofilm infection. Clin. Microbiol. Infect. **19**, 107–112 (2013)
2. Hassett, D.J., Ma, J.F., Elkins, J.G., McDermott, T.R., Ochsner, U.A., West, S.E., Huang, C.T., Fredericks, J., Burnett, S., Stewart, P.S., McFeters, G., Passador, L., Iglewski, B.H.: Quorum sensing in Pseudomonas aeruginosa controls expression of catalase and superoxide dismutase genes and mediates biofilm susceptibility to hydrogen peroxide. Mol. Microbiol. **34**, 1082–1093 (1999)
3. Walker, J.T., Bradshaw, D.J., Fulford, M.R., Marsh, P.D.: Microbiological evaluation of a range of disinfectant products to control mixed-species biofilm contamination in a laboratory model of a dental unit water system. Appl. Environ. Microbiol. **69**, 3327–3332 (2003)
4. Stewart, P.S., Costerton, J.W.: Antibiotic resistance of bacteria in biofilms. Lancet **358**, 135–138 (2001)
5. Macedo, A.J., Abraham, W.-R.: Can infectious biofilms be controlled by blocking bacterial communication? Med. Chem. **5**, 517–528 (2009)
6. Herwaldt, L.A., Cullen, J.J., Scholz, D., French, P., Zimmerman, M.B., Pfaller, M.A., Wenzel, R.P., Perl, T.M.: A prospective study of outcomes, healthcare resource utilization, and costs

associated with postoperative nosocomial infections. Infect. Control Hospital Epidemiol. **27**, 1291–1298 (2006)
7. Darouiche, R.O.: Treatment of infections associated with surgical implants. N. Engl. J. Med. **350**, 1422–1429 (2004)
8. Leone, M., Dillon, L.R.: Catheter outcomes in home infusion. J. Infus. Nurs. **31**, 84–91 (2008)
9. Presterl, E.: Implantatinfektionen. Chemother. J. **18**, 56–60 (2009)
10. Woese, C.R.: Bacterial evolution. Microbiol. Rev. **51**, 221–271 (1987)
11. Fraher, M.H., O'Toole, P.W., Quigley, E.M.: Techniques used to characterize the gut microbiota: a guide for the clinician. Nat. Rev. Gastroenterol. Hepatol. **9**, 312–322 (2012)
12. Schwieger, F., Tebbe, C.C.: A new approach to utilize PCR-single-strand-conformation polymorphism for 16S rRNA gene-based microbial community analysis. Appl. Environ. Microbiol. **64**, 4870–4876 (1998)
13. Pichlmaier, M., Marwitz, V., Kühn, C., Niehaus, M., Klein, G., Bara, C., Haverich, A., Abraham, W.-R.: High prevalence of asymptomatic bacterial colonization of rhythm management devices. Europace **10**, 1067–1072 (2008)
14. Cox, M.J., Cookson, W.O., Moffatt, M.F.: Sequencing the human microbiome in health and disease. Hum. Mol. Genet. **22**(R1), R88–R94 (2013)
15. Pichlmaier, M., Knigina, L., Kuehn, C., Khaladj, N., Oswald, H., Klein, G., Haverich, A., Abraham, W.-R.: The role of cohabitant unusual bacterial species in infection of cardiovascular implantable electronic devices (CIED). Technol. Health Care **21**, 87–96 (2013)
16. Rohacek, M., Weisser, M., Kobza, R., Schoenenberger, A.W., Pfyffer, G.E., Frei, R., Erne, P., Trampuz, A.: Bacterial colonization and infection of electrophysiological cardiac devices detected with sonication and swab culture. Circulation **121**, 1691–1697 (2010)
17. Eribe, E.R., Olsen, I.: Leptotrichia species in human infections. Anaerobe **14**, 131–137 (2008)
18. Ruiter, N.M., Cnossen, T.T., Bakker, R.C., van Keulen, P.J.H.: Sphingobium olei peritonitis: peritoneal dialysis in turmoil? Perit. Dial. Int. **33**, 102–103 (2013)
19. Lee, A., Wang, H.L.: Biofilm related to dental implants. Implant Dent. **19**, 387–393 (2010)
20. Periasamy, S., Kolenbrander, P.E.: Central role of early colonizer Veillonella sp. in establishing multispecies biofilm communities with initial, middle and late colonizers of enamel. J. Bacteriol. **192**, 2965–2972 (2010)
21. Diaz, P.I., Chalmers, N.I., Rickard, A.H., Kong, C., Milburn, C.L., Palmer Jr, R.J., Kolenbrander, P.E.: Molecular characterization of subject-specific oral microflora during initial colonization of enamel. Appl. Environ. Microbiol. **72**, 2837–2848 (2006)
22. Heuer, W., Stiesch, M., Abraham, W.-R.: Microbial diversity of supra- and subgingival biofilms on freshly colonized titanium implant abutments in the human mouth. Eur. J. Clin. Microbiol. Infect. Dis. **30**, 193–200 (2011)
23. Socransky, S.S., Haffajee, A.D., Cugini, M.A., Smith, C., Kent Jr, R.L.: Microbial complexes in subgingival plaque. J. Clin. Periodontol. **25**, 134–144 (1998)
24. Schreiber, S., Rosenstiel, P., Albrecht, M., Hampe, J., Krawczak, M.: Genetics of Crohn disease, an archetypal inflammatory barrier disease. Nat. Rev. Genet. **6**, 376–388 (2005)
25. Burmølle, M., Webb, J.S., Rao, D., Hansen, L.H., Sørensen, S.J., Kjelleberg, S.: Enhanced biofilm formation and increased resistance to antimicrobial agents and bacterial invasion are caused by synergistic interactions in multispecies biofilms. Appl. Environ. Microbiol. **72**, 3916–3923 (2006)
26. Turnbaugh, P.J., Quince, C., Faith, J.J., McHardy, A.C., Yatsunenko, T., Niazi, F., Affourtit, J., Egholm, M., Henrissat, B., Knight, R., Gordon, J.I.: Organismal, genetic, and transcriptional variation in the deeply sequenced gut microbiomes of identical twins. Proc. Natl. Acad. Sci. U.S.A. **107**, 7503–7508 (2010)

Animal Test Models for Implant-Associated Inflammation and Infections

Bushra Rais, Muhammad Imran Rahim, Stefan Lienenklaus, Siegfried Weiss, Christian Tolle, Jan-Marten Seitz, Henning Menzel, Hansjörg Hauser and Peter Paul Müller

Abstract To evaluate the biocompatibility and inflammatory potential of prospective implant materials a mouse model was established using in vivo imaging to monitor inflammatory responses to individual implants over time. Various inflammation associated products and processes were assayed such as reactive oxygen radicals, proteases produced by immune cells, cell stimulatory signaling molecules and interferon gene activation. These were detected either by biochemical activation of fluorescent molecules or by transgenic animals expressing luciferase to monitor inflammatory interferon-β (IFN-β) induction. The results showed that inflammatory signals can be detected by in vivo imaging after subcutaneous implantation of biocompatible or immune stimulatory implants. However, there were specific differences depending upon the assay system. The response to inflammatory proteases and cell growth signaling molecules appeared delocalized and was difficult to assign to one of several implants in individual animals. On the other hand, the interferon response was locally focused and was highly specific for pathogens whereas no signal was detected in response to wounding or to biocompatible implant materials. In conclusion, of the various detection systems investigated, the transgenic interferon mouse model could be applied to monitor bacterial implant infections and will be useful to evaluate the efficacy of antimicrobial implant coatings.

Keywords Implant material · Infection · Inflammation · In vivo imaging · Mouse model · Wound healing · Fluorescence · Bioluminescence

B. Rais · M.I. Rahim · S. Lienenklaus · S. Weiss · H. Hauser · P.P. Müller (✉)
Helmholtz Center for Infection Research, Braunschweig, Germany
e-mail: pmu@gbf.de

C. Tolle · H. Menzel
Technische Universität, Braunschweig, Germany

J.-M. Seitz
Institute of Material Science, Leibniz Universität Hannover, Hannover, Germany

T. Lenarz and P. Wriggers (eds.), *Biomedical Technology*, Lecture Notes
in Applied and Computational Mechanics 74, DOI 10.1007/978-3-319-10981-7_12

1 Introduction

Cell culture test conditions are comparatively quick, well defined and reproducible. Cell morphology and number can be used to evaluate cell compatibility of the material surface [1–3]. However, in vitro results cannot reliably be extrapolated to the in vivo situation. An important difference is for example, the presence of the immune system in vivo. Therefore, it is essential to test novel implant materials in vivo before clinical applications can be envisioned. This includes the evaluation of the tissue compatibility of the materials per se, the processing and surface structures and more recently, potential coatings that may be used for drug release purposes to improve the implant performance [4]. Mainly, such tests characterize the effects on the tissue surrounding the implant and the potential to induce immune reactions. For the detailed analysis of the tissue response to novel implant materials, histology can be used. Simple staining procedures can be used to visualize tissue damage, identify immune cells and the extent of granulation tissue is an indication for the degree of inflammation. Subsequently, the progress of the wound-healing reaction and the formation of fibrous capsule around implants can be evaluated. Staining with antibodies can be used for the highly specific detection of cells, including their differentiation and activation status. The major disadvantage of histological analysis is that solely a single predetermined time point can be analyzed for each implant and each animal. Alternatively, gene expression analysis has become a routine tool for obtaining detailed information about molecular events of the implant tissue interaction. This may serve to identify specific biological mechanisms and find ways to enhance this interaction or to improve the healing process. On the other hand, gene expression analysis generally gives only an average summary over processes in the tissue from many diverse cell types and complex interactions that take place between them. Gene expression analysis is an intensive method and similar to histological analysis, allows a single time point per implant or per animal [5].These methods preclude the monitoring of tissue responses to individual implants over time. Alternatively, in vivo imaging can be used advantageously to quantify and monitor tissue responses over an extended period of time in individual animals [6–12]. In this study, various clinically established implant materials were used for comparison. As biocompatible standard materials titanium and glass-ceramic implants were used. These are clinically used as dental, orthopedic or middle ear implants respectively [13–16]. In addition, inactivated bacteria were used as highly inflammatory and immune-stimulatory agents that could reflect implant infections.

As soon as an implant material comes into contact with body liquids such as blood or interstitial fluid, proteins are adsorbed on the implant surface [17]. It is however not clear to what extend this protein layer affects the subsequent interactions. In addition, inflammatory molecules are released by stressed or damaged cells and by reactions taking place during blood coagulation [18]. These molecules form a concentration gradient that recruits inflammatory cells from the blood circulation. In response, endothelial cells that line nearby blood vessels express adhesion molecules on the cell surface that allow the adhesion of circulating granulocytes and macrophages. These attach loosely, then adhere more firmly to the vessel walls and eventually penetrate them, guided by the concentration gradient they move through the tissue to the site of injury. The migration through the tissue requires the action of inflammatory proteases that degrade extracellular matrix components and lead to the loosening of cell–cell interactions. Polymorphonuclear neutrophils (PMNs) are the first cells that accumulate in the tissue within the first four hours after injury. These cells can recognize danger signals released from injured tissue via various pattern recognition receptors (PRRs). The signaling of these receptors can activate several inflammatory signaling pathways, there by promoting further inflammatory events and the recruitment of more cells from the blood circulation. Activated PMNs produce reactive oxygen species and proteases [19–22]. These reactions are a defense against foreign bodies which, depending on their nature may be degraded and superficial foreign bodies may be extruded. Any pathogens associated with the intrusion of a foreign body are attacked and, if successful, eradicated or at least locally contained until the adaptive immune system takes over. During excessive inflammation, these reactions, even lead to cell death and damage of the surrounding tissue. After a few hours monocytes are attracted from the blood circulation by the action of small peptides called chemokines that are released from PMNs and to a lesser degree by other cells. The monocytes mature to macrophages that control subsequent steps in the tissue, including the wound-healing reaction and the formation of a fibrotic capsule around the foreign body. Thereby, the inflammation is switched from the acute to the chronic stage (Fig. 1) [19, 23]. To establish a reliable testing system that can evaluate the inflammatory potential of novel implant materials we evaluated several strategies to visualize biomaterial-associated inflammatory reactions in vivo. Reactive oxygen species are produced by immune cells in response to inflammation. The oxidizing potential of these radicals was assessed. Hydrocyanines were used as chemical sensors which are oxidized to fluorescent cyanines in the presence of ROS. Cyanines can be visualized after excitation by their fluorescence in the near infra-red spectrum [24]. Alternatively, inflammatory proteases like cathepsins are produced by activated neutrophils or macrophages. The protease activity can be imaged using a specific probe that is hydrolyzed by cathepsins into a fluorescent product [25–27]. In a third approach, the activity of the extracellular lipase autotaxin was used to visualize an inflammatory process. Autotaxin activity generates a lipid signaling molecule that stimulates cell proliferation, cell migration and cell survival [29] and has been shown to be produced during lung inflammation [28]. For imaging, a fluorophore linked to a quencher by an autotaxin sensitive substrate was used. Upon cleavage with autotaxin, the fluorophore and its quencher are separated resulting in increased fluorescence efficiency. In addition, particularly in the early stages during the wound healing processes, implants are prone to colonization by bacteria. Bacterial infection induces cytokines like interferon-β (IFN-β). This has originally been discovered as part of the antiviral response but it is also induced by diverse bacteria [30, 31]. Bacterial infections could be visualized by using a transgenic mouse model in which the interferon(IFN)-β gene is replaced by a luciferase reporter gene that can be used for imaging purposes [32]. For imaging, heterozygous mice were used to allow IFN-β production from the wild type allele. The suitability of these in vivo

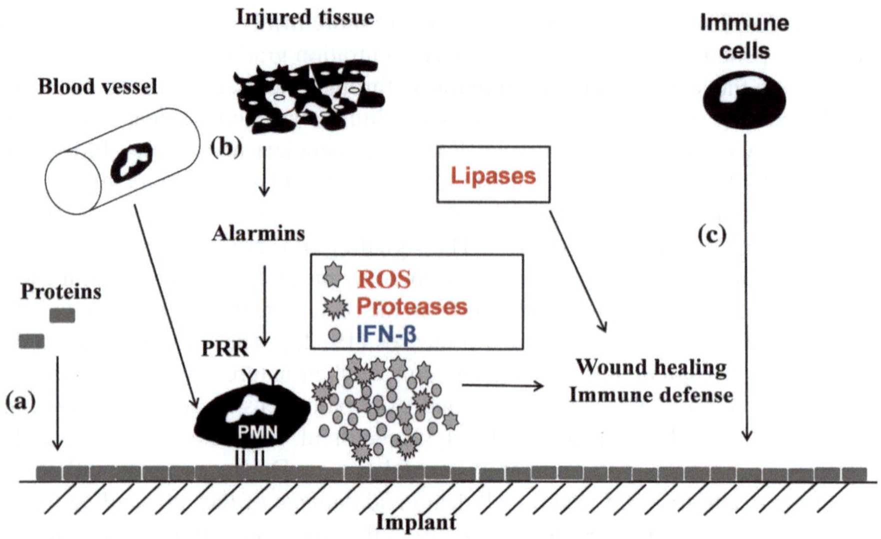

Fig. 1 Implant-associated inflammation. Soon after implantation a protein layer is adsorbed on the implant surface (**a**) Recruitment of PMNs to the site of injury starts following injury and protein adsorption (**b**) Macrophages and monocytes are recruited at later stages (**c**)

imaging approaches was compared for the reliable evaluation and ranking of the biocompatibility of various implant materials and for detecting bacterial infections events in real time, respectively.

2 Materials and Methods

2.1 Preparation of Hydrocyanine

To generate a chemical sensor for oxidative compounds for in vivo imaging 2 mg of indocyanine green (I2633, Sigma Aldrich) was dissolved in 4 ml of methanol (J.T. Baker®, Germany). 3 mg of sodium borohydride (Sigma Aldrich) was added to carry out reduction reaction. The solution was stirred continuously for 5 mins in absence of oxygen. Solvent was evaporated in the presence of vacuum using rotary evaporators. The dried powder was stored at −20 °C overnight. Before injection, the reduced hydrocyanine powder was dissolved in 2 ml of deionized water making up the final concentration to 1 mg/ml [24].

2.2 Heat Inactivation of Staphylococcus Aureus

As a potent inflammation-inducing agent and as a substitute for infectious bacteria heat inactivated *Staphylococcus aureus* suspensions were prepared. The bacteria were streaked out on a Lysogeny broth medium (LB) agarose plate and incubated over night at 37 °C. Single colonies were picked with a sterile needle and used to inoculate a liquid LB culture that was incubated on a rotary shaker at 180 rpm at 37 °C. When the density of the culture reached an OD_{600} of 0.1, 1 ml of the bacterial culture was centrifuged at maximum speed in an Eppendorf centrifuge for 5 mins at room temperature. The supernatant was discarded and the pellet was suspended in 1 ml of phosphate buffered saline (PBS), pH 7.0. For inactivation, the bacteria were first heated to 75 °C for 15 mins and then stored on ice.

2.3 Implant Preparation

For biocompatible and inflammatory implant preparation, respectively, plain material samples or samples coated with bacterial products were used. Porous implants were used to increase the carrier capacity and the stability of the coatings to prolong the release after implantation. Porous glass beads obtained from VitraPOR, Germany (Size 4 mm, Pore size-60 µm) were used as biocompatible implants. Inflammatory porous glass implants were prepared by soaking the beads for 2 min in heat inactivated *Staphylococcus aureus* suspensions and left to dry under ambient conditions. Porous titanium discs of 7 mm diameter and 2 mm thickness were prepared from micro-beads by an injection molding and sintering procedure. Magnesium discs with a diameter of 5 mm and height of 2 mm were prepared by extrusion of a rod followed by cutting off individual discs. Poly-L-lactic acid beads with a diameter of 5 mm were purchased from Good Fellow, England.

2.4 Subcutaneous Implantations in Mice

Wild-type BALB/c mice were obtained from Harlan–Winkelmann laboratories, Germany. Animals were housed under pathogen free conditions in a group of maximum five animals per cage. Mice were anesthetized by intraperitoneal injection of ketamine (10 mg/kg) and xylazine (4 mg/kg). The back was shaved using an electronic razor (Aesculap, Germany). An incision of 1 cm was made in the dorsal skin and a small pouch was made under the skin to insert the implant. The wound was closed by interrupted suturing using polyglactin filaments (Ethicon, Germany). Mock implantation was done by following the complete surgical procedure but without inserting any implant. All animal experiments were done in accordance with the regulations and with the approval from the local authorities Lower Saxony State Office for Consumer Protection and Food Safety (LAVES), permission number 33.42502/07-10.5.

2.5 In Vivo Imaging of the Oxidation Potential

After anesthesia 30 μg of hydrocyanine solution was injected subcutaneously at the site of implantation. After 30 mins, fluorescent imaging was done in the near infrared spectrum using in vivo imaging system (IVIS200, Xenogen, USA). The excitation wavelength of hydrocyanines was 750 nm and the emission wavelength was 840 nm. Acquired images were corrected for the background using image math tool of living image software (version 4.3.1, Caliper life Sciences, 2012. In order to correct for the background, a region of interest was selected from a mouse without implants after addition of the fluorophore. The background value was calculated and automatically subtracted from the images by the software.

2.6 Monitoring Inflammatory Protease Activity

Protease activity dependent fluorescent sensor, Prosense 680 (PerkinElmer, Germany) was used according to the manufacturer's recommendations. A dose of 2 nm was injected intravenously into the mouse tail. Fluorescent imaging of the tissue with implants was done after a period of 30 min with an excitation and emission wavelength of 680 and 700 nm, respectively, using in vivo imaging system (IVIS 200). Similarly, the background of fluorescent images was subtracted using image math tool of living image software as described above.

2.7 Visualization of Cell Growth Signaling

Lipase activity in the mouse tissue was monitored using ATX Red (L2010, Echelon Biosciences Salt Lake City, UT). 10 μg of ATX Red fluorophore was injected intravenously into the mouse tail. Whole animal imaging was done in the near infrared spectrum at an excitation and emission wavelength of 775 and 800 nm, respectively, 30 mins after injection of ATX Red. Images were processed in the same way as explained above.

2.8 Imaging Interferon-β Induction in Transgenic Reporter Mice

Transgenic female BALB/c mice with a luciferase gene replacing the IFN-β coding sequence on one allele were used for non-invasive in vivo imaging of the interferon-β induction [32]. 150 μl of (30 mg/ml) luciferin (PerkinElmer) was injected intraperitoneally. After 15 min, luminescent imaging was performed using an in vivo imaging system (IVIS 200).

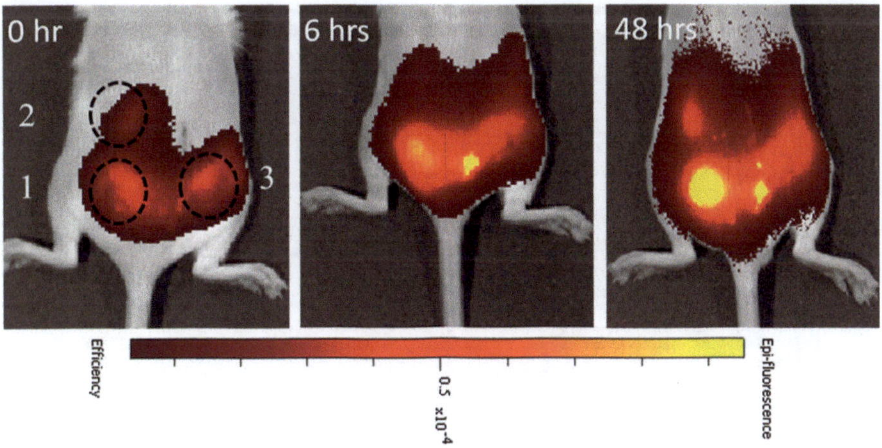

Fig. 2 Enhanced oxidizing potential in response to inflammatory implants visualized by a fluorescent probe. Biocompatible glass implants and glass implants coated with inflammation-inducing heat inactivated *S. aureus* were subcutaneously implanted into Balb/c mice. Hydrocyanine solutions were injected at the site of implantation as oxidation-sensitive dye precursor molecules. Fluorescence intensity was determined at 0, 6 and 48 h after dye injection. The following implants were used: *1* Glass bead coated with inactivated bacteria, *2* Glass implant, *3* Surgery without implant. *Dashed circles* represent the site of implantation

3 Results

3.1 Imaging of Inflammatory Oxygen Radicals

To visualize and compare inflammatory reactions over the time, biocompatible and inflammatory material samples were implanted subcutaneously in mice. Reactive oxygen products of neutrophil and macrophage cells were detected after subcutaneous administration of hydrocyanines. These reduced dyes become fluorescent after an oxidation reaction with oxygen radicals. As highly inflammatory material porous glass beads loaded with heat-inactivated bacteria were used. Fluorescence intensity increased during first two days (Fig. 2). The site with the bacteria appeared brighter fluorescent when compared to bare glass implants. Therefore, oxidation sensitive dyes could be used to differentiate inflammatory and biocompatible implants. However, the response to a biocompatible implant and the wound healing response showed similar fluorescence intensities and could not be differentiated by this method (Fig. 2).

3.2 Imaging Protease Activity in Inflammation

Proteases are secreted by immune cells and essential to allow migration through the tissue. To image inflammatory protease activity in vivo, inflammatory and

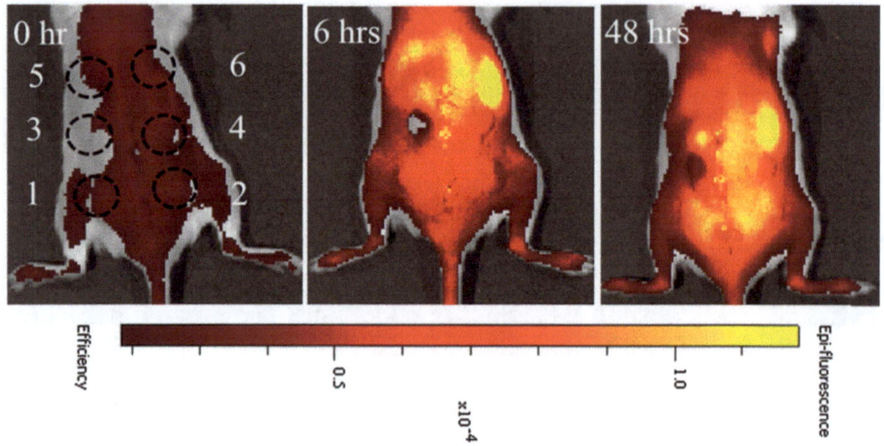

Fig. 3 Delocalized protease activity in response to pro inflammatory implants over the time period. Fluorescent images were taken at 0, 6 and 48 h. The format of implantation was following: *1* Porous glass implant soaked in dead bacteria, *2* Porous glass implant, *3* Magnesium, *4* Poly-L-lactic acid, *5* Heat killed bacteria and *6* Surgery without implantation. *Dashed circles* represent the site of implantation

non-inflammatory materials were subcutaneously implanted in mice. A fluorophore that can be activated by inflammatory proteases (cathepsins) was injected intravenously and whole body imaging was performed. The results showed an increase in fluorescence intensity with time (Fig. 3). The signal peaks appeared to be delocalized and could not be assigned to individual implants. Furthermore, opaque implants quenched even the background fluorescence. This phenomenon could be used to precisely localize the position of the implants (Fig. 3; implant 3 at 6 h and 48 h). At 48 h, the implant with bacteria showed brighter fluorescence intensity as compared to biocompatible glass implants (Fig. 3; 48 h).

3.3 Imaging Cell Growth Stimulatory Signals

Injury stimulates cell growth and tissue repair processes. Cell growth can be stimulated by lipase signaling [29]. Autotaxin is a specific lipase involved in cell growth and motility signaling. ATX Red is a near infra-red fluorescent precursor molecule that can act as autotaxin lipase substrate. A fluorescent product is generated by autotaxin that can be detected by in vivo imaging. Various biocompatible and inflammatory implant materials were implanted in mice to evaluate this method. After implantation, mice were injected intravenously with ATX red (Fig. 4a, b, 0 h). To monitor the basal levels of autotaxin, one mouse without implants was included (Fig. 4c). Implants with bacteria induced a higher autotaxin activity as compared to biocompatible glass implant (Fig. 4; implant 2, 48 h). A signal peak was obtained 48 h after

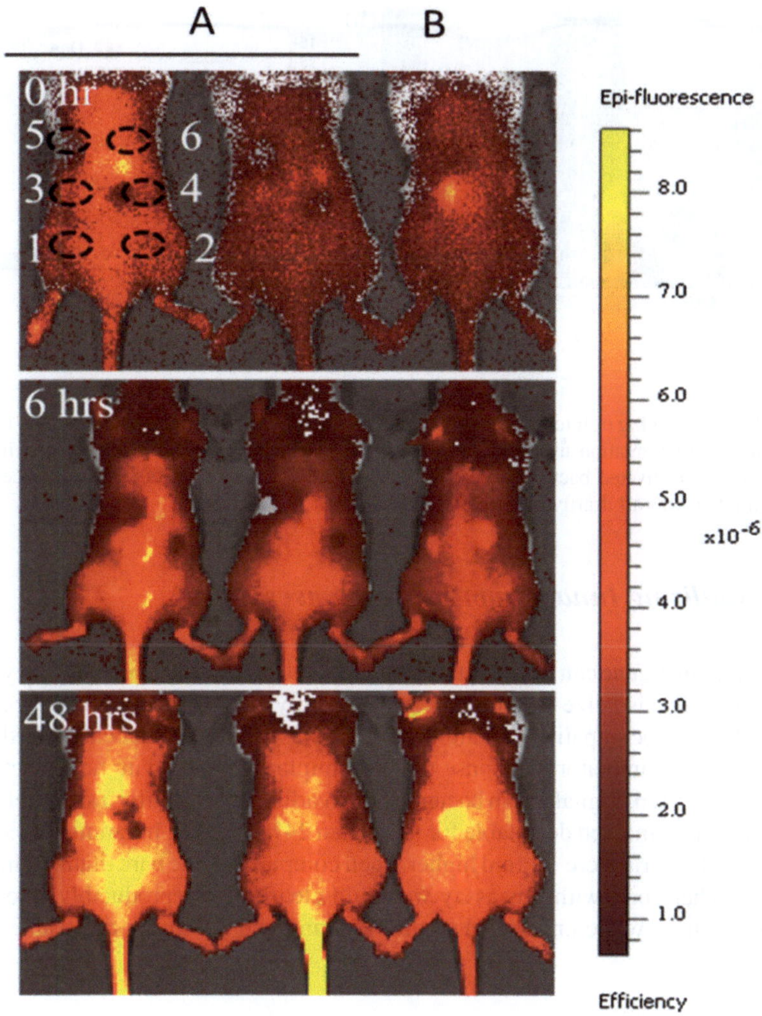

Fig. 4 Imaging of increased but delocalized inflammatory cell proliferation signals. Mice were implanted with various inflammatory and non-inflammatory materials. ATX-Red was applied and fluorescent images of the following time points: 0, 6 and 48 h. First two mice from left were with implants and the third mouse is without implants. Implants were as follows: *1* Porous glass implant, *2* Porous glass implant with dead bacteria, *3* Poly-L-lactic acid, *4* Magnesium, *5* Mock and *6* Titanium. *Dashed circles* represent the site of implantation

the dye injection at a site where there was no implant and which may correspond to the location of the spleen (Fig. 4, 48 h, yellow spot on the left side of each animal). However, there was a high background signal and the inflammatory signals were delocalized, making it difficult to determine the response to individual implants.

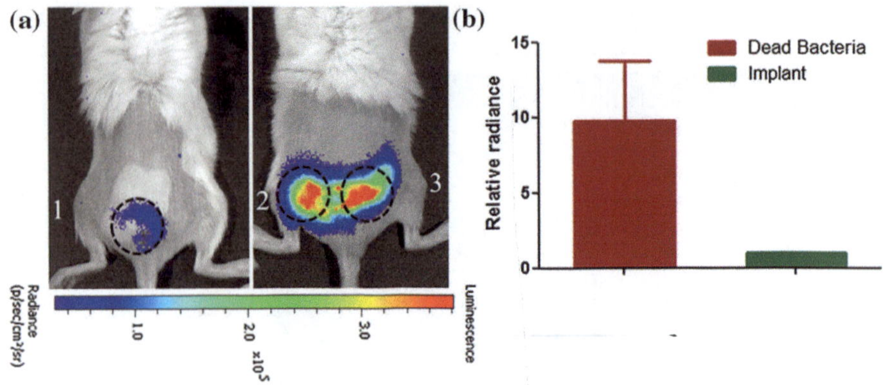

Fig. 5 In vivo imaging of Interferon-β production in response to bacteria: Bioluminescent imaging was done after implantation and luciferin injection (**a**). Implants are as follows: *1* Titanium implant, *2* and *3* Heat inactivated bacteria. The relative radiance was calculated and the radiance of the control implant was arbitrarily set to 1 (**b**)

3.4 Visualizing Innate Immune Responses

The interferon-β gene can be activated by various infectious pathogens such as viruses or bacteria. To visualize interferon-β gene activity transgenic reporter mice were used [32]. A biocompatible titanium implant was implanted subcutaneously and to induce an inflammatory response titanium implants coated with heat inactivated bacteria were used. Signals from bacteria were readily detected after a short delay of a few hours and remained detectable for more than 24 h (Fig. 5a). Inflammatory signals induced by bacteria were 10-fold higher when compared to the bare titanium implant (Fig. 5b). Therefore, with this assay we could clearly and reproducibly detect and localize bacteria while uninfected control implants did not give a detetcable signal (Fig. 5a).

4 Discussion

We have attempted to establish a method to distinguish and evaluate the inflammatory potential of implant materials. For this purpose, we used different fluorescent reporter molecules and luminescent transgenic reporter animals. It was first attempted to apply hydrocyanines intravenously but the resulting fluorescence intensity was below the detection limit (data not shown). Therefore, the dye was applied directly at implantation site. The inflammation associated oxidation potential increased in response to bacterial components over the time period investigated. Implantation of biocompatible glass also induced some minor degree of ROS activity which was comparable to surgery without implantations suggesting a response caused by injury due to surgery rather than due to the implant. In principle, the method would be suitable to detect bacterial infection; however, the fluorescence signal was highly variable in independent experiments. Visualization of inflammatory protease activity also showed increased signal intensity after implantations. However, the signal was delocalized and could not be assigned to individual implants. For our purposes, this method did not appear suitable to distinguish inflammatory potential of different implant materials. The approach to monitor cell growth stimulatory signals has been used to detect tumors and is also expected to be associated with cell proliferation during wound healing processes [33]. Autotaxin activity was observed to increase with time at the site of implants associated with bacteria. Unfortunately the signals were not sufficiently localized to allow any reliable assignment to individual implants. In addition, the signal was barely detectable due to the high background [34]. Therefore, it was not possible to quantify the inflammatory potential of individual implants using this strategy.

Strategies for visualizing enzymatic activities or presence of reactive molecules using fluorescent probes generally suffer from high background due to autofluorescence of tissue as well as background fluorescence of the probe. The transgenic mouse model employed in this study uses bioluminescence to visualize the induction of a IFN-β as an inflammation related cytokine. The high signal to background ratio facilitated the quantification and a distinct difference between bacteria coated and non-inflammatory titanium implants was observed. No interferon response was observed with any other clinically used implant material (data not shown). This assay could therefore be used to investigate bacterial infections on implant materials and to determine the efficacy of antibacterial implant coatings.

The major advantage of in vivo imaging is that each animal can serve to include internal controls by using appropriate materials implanted in the same animal, thereby improving the reproducibility by avoiding even subtle differences in the handling and animal to animal variability [35]. Both, fluorescent as well as bioluminescent strategies could be used.

In conclusion, various techniques were investigated to evaluate the inflammatory potential of prospective implant materials. Compared to the fluorescent detection methods investigated here, a luminescent transgenic mouse model appeared most promising to distinguish biocompatible and immune stimulatory implants.

Acknowledgments This research was funded in part by Collaborative Research Centre grant SFB599 by the Deutsche Forschungsgemeinschaft. We thank Dr. Thomas Ebel from Helmholtz Centre Geesthacht, Germany for providing porous titanium discs. B.R. was funded by Helmholtz graduate school and M.I.R was supported by a joint grant of DAAD (German Academic Exchange Service), Germany and Higher Education Commission of Pakistan.

References

1. Schade, R., et al.: Biomimetic organic-inorganic nanocomposite coatings for titanium implants. in vitro and in vivo biological testing. J. Biomed. Mater. Res. A **95**(3), 691–700 (2010)
2. Sauberlich, S., et al.: Cell culture tests for assessing the tolerance of soft tissue to variously modified titanium surfaces. Clin. Oral. Implants Res. **10**(5), 93–379 (1999)

3. Noth, U., et al.: Standardized testing of bone implant surfaces with an osteoblast cell culture system. II. Titanium surfaces of different degrees of roughness. Biomed. Tech. (Berl) **44**(1–2), 6–11 (1999)
4. Hesse, D., et al.: Layered double hydroxides as efficient drug delivery system of ciprofloxacin in the middle ear: an animal study in rabbits. J. Mater. Sci. Mater. Med. **24**(1), 36–129 (2013)
5. Badar, M., et al.: The formation of an organic coat and the release of corrosion microparticles from metallic magnesium implants. Acta Biomater. **9**(7), 7580–7589 (2013)
6. Hansch, A., et al.: In vivo imaging of experimental arthritis with near-infrared fluorescence. Arthr. Rheum. **50**(3), 7–961 (2004)
7. Wunder, A., et al.: In vivo imaging of protease activity in arthritis: a novel approach for monitoring treatment response. Arthr. Rheum. **50**(8), 65–2459 (2004)
8. Irmler, I.M., et al.: In vivo molecular imaging of experimental joint inflammation by combined (18)F-FDG positron emission tomography and computed tomography. Arthr. Res. Ther. **12**(6), R203 (2010)
9. Ignatius, M.S., et al.: In vivo imaging of tumor-propagating cells, regional tumor heterogeneity, and dynamic cell movements in embryonal rhabdomyosarcoma. Cancer Cell **21**(5), 680–93 (2012)
10. Condeelis, J., Weissleder, R.: In vivo imaging in cancer. Cold Spring Harb. Perspect. Biol. **2**(12), a003848 (2010)
11. Cagnin, A., Gerhard, A., Banati, R.B.: In vivo imaging of neuroinflammation. Eur. Neuropsychopharmacol. **12**(6), 6–581 (2002)
12. Vasudeva, K., et al.: Imaging neuroinflammation *in vivo* in a neuropathic pain rat model with near-infrared fluorescence and 19F magnetic resonance. PLoS ONE **9**(2), e90589 (2014)
13. Affairs, A.C.O.S.: Titanium applications in dentistry. J. Am. Dent. Assoc. **134**(3), 347–349 (2003)
14. Jorge, J., et al.: Titanium in dentistry: historical development, state of the art and future perspectives. J. Indian Prosthodont. Soc. **13**(2), 71–77 (2013)
15. Beleites, E., Gudziol, H., Höland, W.: Maschinell bearbeitbare Glaskeramik für die Kopf-Hals-Chirurgie. HNO-Praxis **13**, 121–125 (1988)
16. Pinkert, R.: Individually produced open endosseous dental implants made of bioverit glass ceramic. first report on experiences. Zahn Mund Kieferheilkd Zentralbl **78**(5), 6–411 (1990)
17. Wilson, C.J., et al.: Mediation of biomaterial-cell interactions by adsorbed proteins: a review. Tissue Eng. **11**(1–2), 1–18 (2005)
18. Müller, P.: Blood Coagul. Fibrinolysis **53**, 357–365 (2013)
19. Kobayashi, S.D., et al.: Neutrophils in the innate immune response. Arch. Immunol. Ther. Exp. **53**(6), 505–517 (2005)
20. Bianchi, M.E.: DAMPs, PAMPs and alarmins: all we need to know about danger. J. Leukoc. Biol. **81**(1), 1–5 (2007)
21. Nimeri, G., et al.: The influence of plasma proteins and platelets on oxygen radical production and F-actin distribution in neutrophils adhering to polymer surfaces. Biomaterials **23**(8), 1785–1795 (2002)
22. Nimeri, G., et al.: Oxygen radical production in neutrophils interacting with platelets and surface-immobilized plasma proteins: Role of tyrosine phosphorylation. J. Biomed. Mater. Res. Part A **67**(2), 439–447 (2003)
23. Yamashiro, S., et al.: Phenotypic and functional change of cytokine-activated neutrophils: inflammatory neutrophils are heterogeneous and enhance adaptive immune responses. J. Leukocyte Biol. **69**(5), 698–704 (2001)
24. Selvam, S., et al.: Minimally invasive, longitudinal monitoring of biomaterial-associated inflammation by fluorescence imaging. Biomaterials **32**(31), 7785–92 (2011)
25. Bratlie, K.M., et al.: Rapid biocompatibility analysis of materials via *in vivo* fluorescence imaging of mouse models. PLoS ONE **5**(4), e10032 (2010)
26. Faurschou, M., Borregaard, N.: Neutrophil granules and secretory vesicles in inflammation. Microbes Infect. **5**(14), 27–1317 (2003)

27. Lominadze, G., et al.: Proteomic analysis of human neutrophil granules. Mol. Cell. Proteomics **4**(10), 21–1503 (2005)
28. Park, G.Y., et al.: Autotaxin production of lysophosphatidic acid mediates allergic asthmatic inflammation. Am. J. Respir. Crit. Care Med. **188**(8), 40–928 (2013)
29. Perrakis, A., Moolenaar, W.H.: Autotaxin: structure-function and signaling. J. Lipid Res. **55**(6), 1010–1018 (2014)
30. Decker, T., Muller, M., Stockinger, S.: The yin and yang of type I interferon activity in bacterial infection. Nat. Rev. Immunol. **5**(9), 87–675 (2005)
31. Trinchieri, G.: Type I interferon: friend or foe? J. Exp. Med. **207**(10), 2053–2063 (2010)
32. Lienenklaus, S., et al.: Novel reporter mouse reveals constitutive and inflammatory expression of IFN-beta in vivo. J. Immunol. **183**(5), 36–3229 (2009)
33. Madan, D., et al.: Non-invasive imaging of tumors by monitoring autotaxin activity using an enzyme-activated near-infrared fluorogenic substrate. PLoS ONE **8**(11), e79065 (2013)
34. Soubret, A., Ntziachristos, V.: Fluorescence molecular tomography in the presence of background fluorescence. Phys. Med. Biol. **51**(16), 3983 (2006)
35. Sjollema, J., et al.: The potential for bio-optical imaging of biomaterial-associated infection in vivo. Biomaterials **31**(8), 1984–1995 (2010)

MIX
Papier aus verantwortungsvollen Quellen
Paper from responsible sources
FSC® C105338

If you have any concerns about our products,
you can contact us on
ProductSafety@springernature.com

In case Publisher is established outside the EU,
the EU authorized representative is:
Springer Nature Customer Service Center GmbH
Europaplatz 3, 69115 Heidelberg, Germany

Printed by Libri Plureos GmbH
in Hamburg, Germany